Haptic Visions

Visual Rhetoric
Series Editor: Marguerite Helmers

The Visual Rhetoric series publishes work by scholars in a wide variety of disciplines, including art theory, anthropology, rhetoric, cultural studies, psychology, and media studies.

Other Books in the Series

Locating Visual-Material Rhetorics: The Map, the Mill, and the GPS by Amy D. Propen (2012)

Visual Rhetoric and the Eloquence of Design, ed. by Leslie Atzmon (2011)

Writing the Visual: A Practical Guide for Teachers of Composition and Communication, ed. by Carol David and Anne R. Richards (2008)

Ways of Seeing, Ways of Speaking: The Integration of Rhetoric and Vision in Constructing the Real, ed. by Kristie S. Fleckenstein, Sue Hum, and Linda T. Calendrillo (2007)

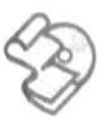

HAPTIC VISIONS

Rhetorics of the Digital Image, Information, and Nanotechnology

Valerie L. Hanson

Parlor Press
Anderson, South Carolina
www.parlorpress.com

Parlor Press LLC, Anderson, South Carolina, USA

S A N: 2 5 4 - 8 8 7 9

Library of Congress Cataloging-in-Publication Data

Hanson, Valerie L., 1969-
Haptic visions : rhetorics of the digital image, information, and nanotechnology / Valerie L. Hanson.
pages cm. -- (Visual rhetoric)
Includes bibliographical references and index.
ISBN 978-1-60235-550-7 (pbk. : acid-free paper) -- ISBN 978-1-60235-551-4 (hardcover : acid-free paper)
1. Technical literature--Philosophy. 2. Visual communication. 3. English language--Rhetoric. 4. Nanotechnolgy. 5. Nanoart. 6. Haptic devices. 7. Scanning tunneling microscopy. I. Title.
T11.H275 2015
601'.4--dc23

2015006894

2 3 4 5

Visual Rhetoric Series
Editor: Marguerite Helmers

Cover design by Danielle Shuff.
Cover image: "Swirl" by Vladimir Kramer. From Unsplash. Used by permission.
Printed on acid-free paper.

Parlor Press, LLC is an independent publisher of scholarly and trade titles in print and multimedia formats. This book is available in paper, cloth and eBook formats from Parlor Press on the World Wide Web at http://www.parlorpress.com or through online and brick-and-mortar bookstores. For submission information or to find out about Parlor Press publications, write to Parlor Press, 3015 Brackenberry Drive, Anderson, South Carolina, 29621, or email editor@parlorpress.com.

Contents

Illustrations

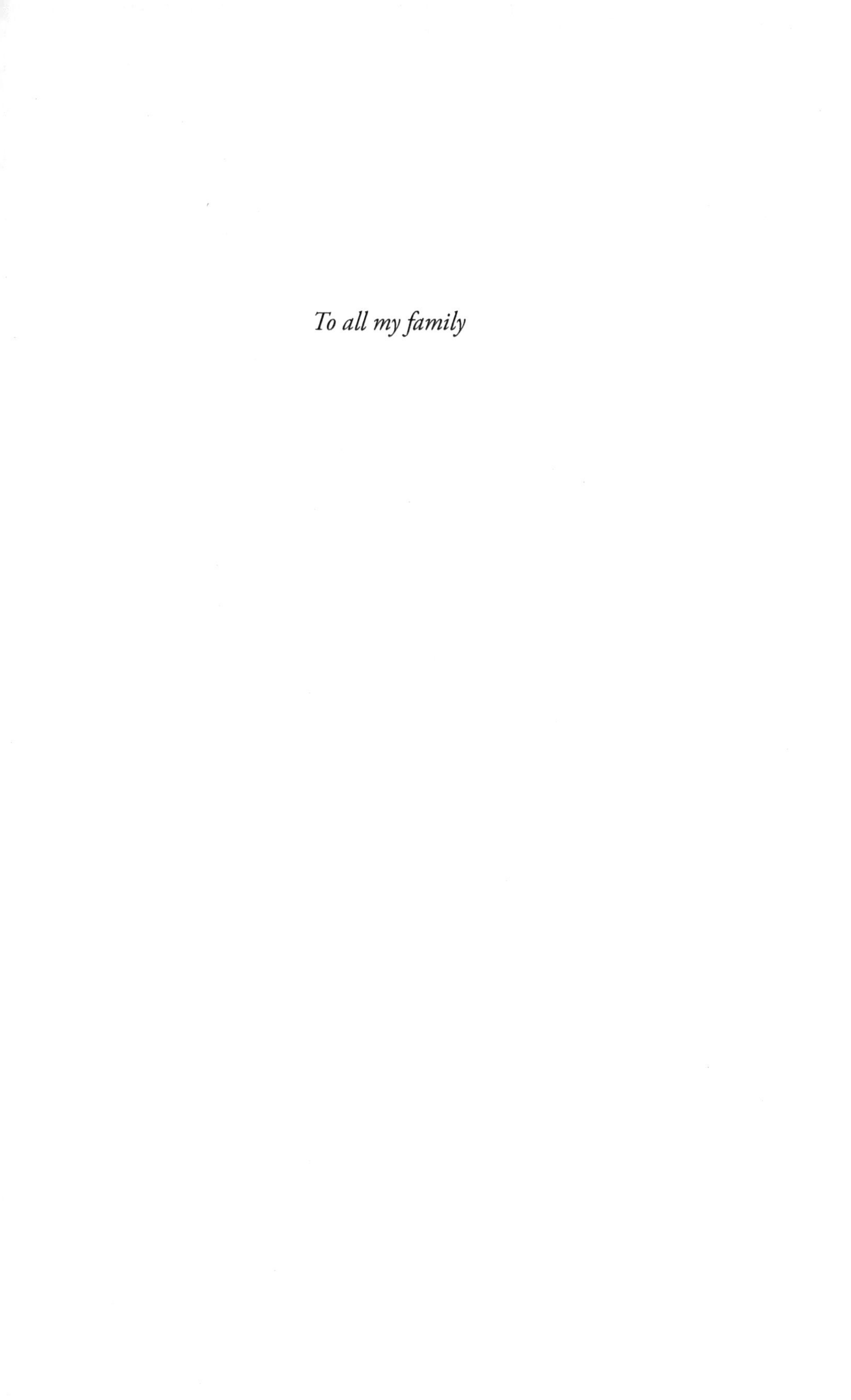

To all my family

Acknowledgments

This book has been a project long in the making. The ideas for this book originally started in my PhD dissertation work at the Pennsylvania State University; I am grateful to my friends, my teachers, and my dissertation committee—Richard Doyle, Stuart Selber, Susan Squier, and Robert Yarber—for their inspiring and enlightening conversations, provocations, insights, and advice. I am also grateful for graduate fellowships from Penn State and from a National Science Foundation grant (Science, Medicine, and Technology in Culture; Pennsylvania State University, 2002–2003; principal investigators: Londa Schiebinger, Robert Proctor, Richard Doyle, and Susan Squier) that gave me time to pursue these ideas.

The collegial and intellectual support I have received at Philadelphia University helped me continue this project. In particular, I want to thank Marion Roydhouse, Katharine Jones, John Eliason, and Julie Kimmel for their encouragement and conversation. A Philadelphia University Research and Design grant helped expand my research through funding interviews conducted with scientists using the scanning tunneling microscope (STM) during 2005 and 2006. I am deeply grateful for the generosity, time, and good will of those scientists who agreed to be interviewed for this project. Thanks also go to the members of the Philadelphia-area nano-studies reading group for their intellectual support, insights, and discussions of things nano and beyond, including a few of this book's topics in earlier stages.

A slightly modified form of parts of Chapter 4 and the conclusion originally appeared in *Science Communication* ("Amidst Nanotechnology's Molecular Landscapes: The Changing Trope of Subvisible Worlds," *Science Communication*, 34.1 (2012): 57–83. Pre-published May 19, 2011 (DOI: 10.1177/1075547011401630)). I am grateful to the editor, Susanna Hornig Priest, and the anonymous reviewers for their guidance. An earlier version of the argument in Chapter 3 appeared in a different form in *Augenblick* ("Nature as Database? Microscope Images' Impact on Visual Cultures of the Natural World," *Augenblick*, 45 (2009): 9–25). I thank the guest editor, Angela Krewani, for inviting me to be part of that production. This book's development also benefited from a fellowship as part of the ZiF Research Group, "Science in the Context of Application: Methodological Change, Conceptual Transformation, Cultural Reorientation," at Zentrum für interdisziplinäre Forschung (Center for Interdisciplinary Research), and I thank the other fellows and organizers of that fellowship. A few ideas developed in Chapter 3 appear in a different, earlier form in in my contribution to *Science Transformed? Debating Claims of an Epochal Break,* edited by Alfred Nordmann, Hans Radder and Gregor Schiemann (University of Pittsburgh Press, 2011). Parts of this book's arguments were also presented at conferences and other presentations, including those at the ZiF, the Annual Meetings of the Society for the Social Studies of Science (4S), the European Association for the Study of Science and Technology (EASST) Conferences, Conferences on College Composition and Communication (CCCCs), and Imaging and Imagining NanoScience and Engineering: An International and Interdisciplinary Conference. I thank audience members for their feedback and responses. I also am deeply grateful to David Blakesley and Marguerite Helmers at Parlor Press for their feedback and support, and to the anonymous reviewer whose comments and suggestions markedly improved this book.

Other support, material and emotional, was provided by so many to whom I am so grateful: A tremendous thank you to all my family and friends for conversation, patience, love, understanding, and enthusiasm as this project developed.

Haptic Visions

Introduction: Imaging and Imagining Science in the Information Age

The Information Age incarnates itself in the eye.

—Ivan Illich

In Greg Bear's 1985 and 1988 science fiction novels, *Eon* and *Eternity,* humans from the distant future communicate through a mix of speaking, gesturing, and "picting," where the communicator projects stylized images above her or his shoulder from a torque-shaped machine worn around the neck. In Bear's narrative, this multi-modal communication surprises humans from the near future as they encounter their distant descendants; the use of "picting" while speaking likely seemed far-fetched to Bear's readers in the 1980s. However, as I write this book in the early twenty-first century, with my camera phone in my pocket and my laptop with its graphic user interface on my desk, the use of images to communicate has become ubiquitous. In fact, Bear's envisioned form of communication seems not only plausible today, but also imminent—just a small step from texting with emoji or using Snapchat.

A few years after the publication of Bear's science fiction novels, two scientists, D. M. Eigler and E. K. Schweizer, published a series of six images (see Figure 1) in an article in the April 5, 1990 issue of *Nature.* Like Bear's form of communication, Eigler and Schweizer's images intermingle text, picture, and (atomic) bodily movement to communicate; in the case of Eigler and Schweizer's images, the purpose is to demonstrate the scientist's ability to manipulate thirty-five individual xenon atoms and arrange the atoms to form the letters "IBM" on a nickel surface. At the time, Eigler and Schweizer's images created a stir among scientists and others; the images also spurred interest in an emerging field of science and technology: nanotechnology.[1] The fact that Eigler and Schweizer communicate what we can see and do with atoms in images forms part of a larger message about the important role of visualizations in shaping and communicating scientific knowledge. While the "IBM" images were published a few decades ago, the message that nanoscale images such as Eigler and Schweizer's sends is still relevant today to both scientific and non-scientific audiences. The "IBM" images communicate—and communicate persuasively—about nanotechnology: as the images do so, the "IBM" images affect both the content and function of scientific discourses. Further, Eigler and Schweizer's images form a part of the broader cultural trends in imaging occasioned by the widespread adoption of digital images to communicate—trends that make Bear's vision of communication seem right around the corner.

Haptic Visions is about reading messages conveyed about the nanoscale and image use generally, with a particular focus on the rhetorical interactions among images, ourselves, and the material world. More specifically, this book explores how visualizations like Eigler and Schweizer's form persuasive elements in arguments about manipulation and interaction at the atomic scale. *Haptic Visions* also analyzes how arguments about atomic interaction expressed in images of the nanoscale affect our understanding of nanotechnology, as well as what visualizations like the "IBM" images imply about how digital images and scientific visualization technologies, such as the one that Eigler and Schweizer used (the scanning tunneling microscope or STM), help constitute arguments. While digital imaging and nanotechnology are relatively new developments, what is significant for rhetoricians is not the newness of these two developments, per se. Instead, what is important is that the conventions, practices, and even the significance of both digital imaging and nanotechnology are still in flux. The intersection of digital imaging and nanotechnology thus becomes a site for exploring what becomes persuasive within a developing technology of visual communication—and how then that persuasive communication affects developing scientific and technical fields like nanotechnology.

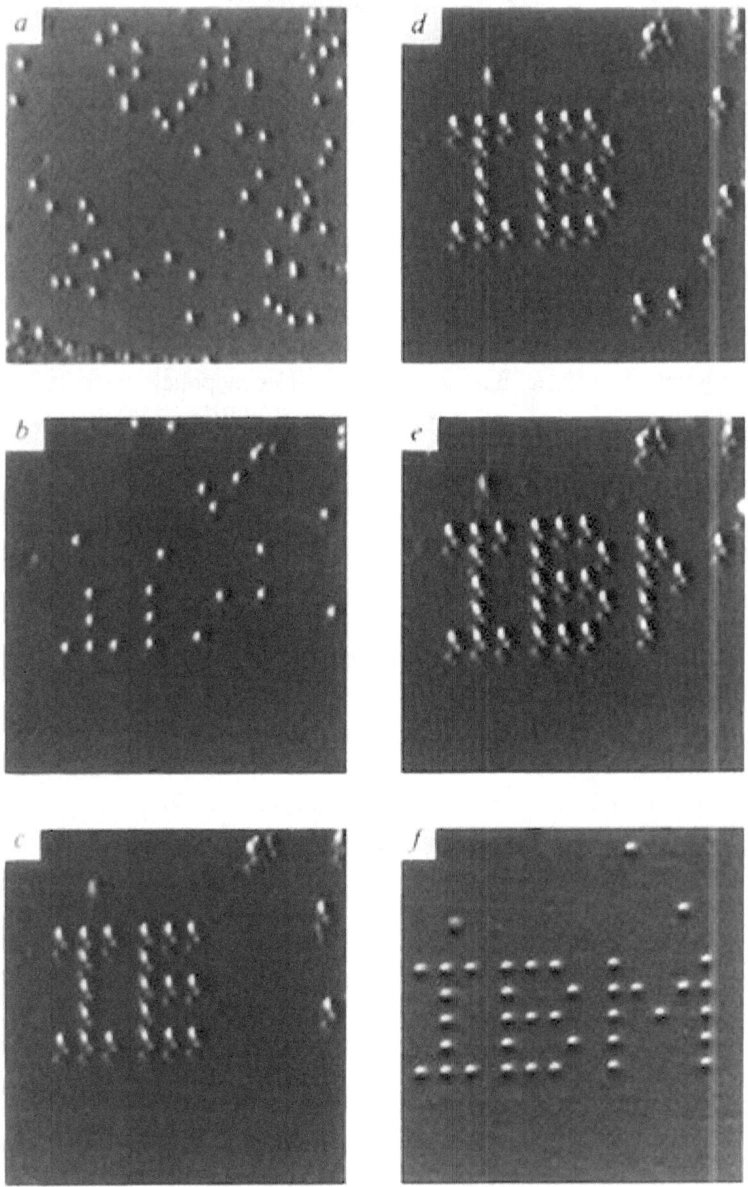

Figure 1. Six images showing the assembly of the letters I, B, and M. Reprinted by permission from Macmillan Publishers, Ltd.: *Nature* (Eigler, D. M. and E. K. Schweizer, "Positioning Single Atoms with a Scanning Tunneling Microscope," *Nature* 344 (April 1990): 525); © 1990.

Why Nanotechnology?

What and how the "IBM" images argue about nanotechnology is important for rhetoric as well as for science: Eigler and Schweizer's demonstration spurred not only excitement and wild visions of the future,[2] but also arguments for serious and funded research in nanotechnology,[3] an emerging field in which researchers from various disciplines—including physics, chemistry, and engineering—study and manipulate phenomena at the nanoscale, the level of single atoms and small molecules.[4] Indeed, the demonstration of atomic manipulation embedded in the "IBM" images forms a topos, or commonplace, in arguments justifying nanotechnology. Science policy justifications for the establishment and funding of this new field,[5] as well as arguments directed towards a popular audience, mention Eigler and Schweizer's demonstration. As historians of science Cyrus C. M. Mody and Michael Lynch observe, while a small proportion of the work in nanotechnology involves manipulating individual atoms, some of the examples of manipulation have become well-known (Mody and Lynch 431, note 19). Demonstrations of atomic manipulation have continued. For example, in May 2013, IBM released the stop-action film, *A Boy and His Atom*, as an advertisement for the company's atomic data storage capabilities. To make the film, which shows a stick-figure boy playing with a ball and then jumping on a trampoline, a research team arranged individual carbon monoxide molecules to form the boy, the ball, and the trampoline, as well as the words "think" and "IBM," all illustrating the researchers' sustained ability for precise manipulation (*A Boy and His Atom*). While nanolithography and related techniques may comprise much of the actual work of nanotechnology, as opposed to the manipulation of individual atoms (Mody and Lynch 431), the idea of atomic manipulation remains a potent, continuing element in nanotechnology discourses, and so deserves examination.

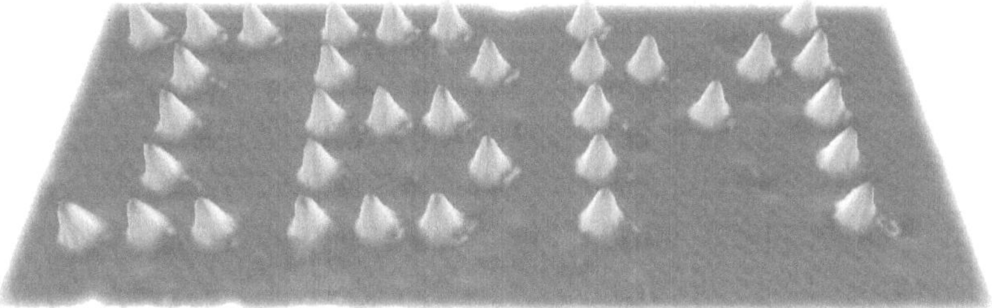

Figure 2. Color image of Eigler and Schweizer's arrangement of xenon atoms into the letters I, B, and M, called "The Beginning" in the STM Image Gallery on the IBM Research web site. Image originally created by IBM Corporation.

How the idea of manipulation and interaction with the nanoscale has lodged into the discourse of nanotechnology so solidly becomes interesting for rhetoric: Following how manipulation and interaction became part of the story of nanotechnology helps to explain not only how a topos forms within a discourse, but also the ways in which rhetorics are formed within—and help form—the complexities of an emerging field. As technology assessment scholar and physicist Ulrich Fiedeler argues, even for an emerging technology, in which stages of development often include discourse, or "communication and negotiation processes" (248), discourse about nanotechnology is a dominant characteristic of the field (246). Therefore, the rhetorics of nanotechnology discourse contribute to the formation of nanotechnology as a field. Fiedeler accounts for the importance of discourse through the interaction of what he sees as the other four main characteristics of nanotechnology: the lack of a clear definition; the interdisciplinarity of nanotechnology research; the fact that nanotechnology is an enabling technology (for example, as with microelectronics, nanotechnology is used to create a variety of other technologies); and the early development stage of nanotechnology products and applications (243). Each of Fiedeler's five main characteristics contribute to the field's complexity, even for those in nanotechnology; researchers and other participants rely more on discourse to bridge disciplines and other forms of knowledge in order to learn about others' research and to conduct their own research in interdisciplinary teams. Further, Fiedeler comments that the sheer number of directions for research and topics for investigation adds to the complexity of nanotechnology, and so also increases the importance of communication (248).

Fiedeler's observation that nanotechnology's discourses form a key characteristic of the field indicates part of nanotechnology's interest for rhetoricians. Nanotechnology becomes a useful site in which to explore how arguments communicate, and in so doing, help constitute a scientific and technical field. How the dominant role of discourse affects nanotechnology and helps shape arguments within the field then become key questions for rhetorical research. While social scientists and rhetoricians have explored some aspects of nanotechnology discourses[6] further analyses of the discursive rhetorics of nanotechnology can illuminate how discourses affect the development of nanotechnology as a field as well as the development of knowledge about the nanoscale.

Rhetorical Roles of Images in Nanotechnology

Images form an important part of nanotechnology discourse, as in any other scientific discourse.[7] Images function within a discourse to constitute as well as communicate knowledge. One important way that images help constitute science is that visualizations focus the attention of viewers on what the images constitute as the main object, or, in science, the scientific object. As Lynch observes, drawing on his considerable and influential ethnographic work on images in science, "instrumental and graphic faculties are implicated in the very organization of what the specimen consists of as a scientific object" ("Externalized Retina" 170). Images then also communicate what a field like nanotechnology establishes as scientific objects. The images used to communicate in a given field also affect that field's formation. As historian of science Martin Rudwick argues in the case of the development of the field of geology, images play a significant role in establishing scientific fields because they help construct "a visual language that [is] appropriate to the subject matter of the science, and which could complement verbal descriptions and theories by communicating observations and ideas that could not be expressed in words" (177).[8] Rudwick and Lynch's comments highlight the importance of analyzing images like Eigler and Schweizer's "IBM" images, because the images function in three ways: they form elements of discourse, they help constitute nanotechnology's scientific objects, and they help establish nanotechnology as a field.

Further, following the rhetorical roles of images can reveal important details about how a different or new visualization technology may influence a field's development. Historians Lorraine Daston and Peter Galison comment that images that show manipulation at the nanoscale, such as images made with the STM and its relative, the atomic force microscope (AFM), differ from others in the history of scientific images through the use of nanoscale images as tools to build objects at the nanoscale (382–84), such as, for example, Eigler and Schweizer's letters in the "IBM" images, or images of the boy, the ball, and the trampoline in *A Boy and His Atom*. Daston and Galison argue that the function of images that allows scientists to construct nanoscale objects alters what images like Eigler and Schweizer's express; images like the "IBM" series do not communicate a representation of nature as much as a presentation—presentation of new objects, wares, and art (47). While Daston and Galison do not present a full analysis of nanoscale images in their history of scientific objectivity, their claims suggest possible connections between new functions of images in nanotechnology and broader trends in visualization and scientific work. Analysis of how manipulation becomes a topos in nanotechnology then may also articulate how images are used in arguments, in terms of the images' functions and messages. Focusing on manipulation and the functions of images together draws a few other important factors into the analysis, including: the visualization technology that Eigler and Schweizer and the makers of *A Boy and His Atom* used, the STM; the imaging processes researchers use to produce STM images; and the vision practices we engage in as we use the STM and other visualization technologies like it. Thus, following the concept of manipulation with attention to image functions also allows us to examine current, cultural visual shifts that affect scientific fields, knowledge, and discourse while exploring the rhetorics of visualization technologies and digital images.

If images and visualization technologies help constitute scientific objects and fields, and if the use of images to manipulate comprises a new function of images, then how the combination of scientific object and new function affect what is depicted in images that are used to communicate nanotechnology becomes a key area for rhetoricians. How the combination of depicting and manipulating atoms affects STM images' uses in discourses—especially those that help to establish nanotechnology—and how STM images are used to persuade within these discourses form two themes of *Haptic Visions.*

Insights about how images like those Eigler and Schweizer published help constitute nanotechnology may also be useful for studies of other disciplines, especially as the social sciences, other sciences, and (increasingly) the humanities rely more on data and computation as part of knowledge-making.[9] As scholars in emerging or established fields manage and communicate massive quantities of data as part of their research, the role images play in communication and knowledge-making strategies becomes more important to understand. Following how images function to communicate—and communicate persuasively—helps articulate the impact on fields and on the rhetorics those fields are composed of and compose.

Digital Data Visualization Rhetorics and Nanotechnology

Eigler and Schweizer's "IBM" images also are significant as examples of productions of recent digital visualization technologies within broader cultural trends of image creation and circulation that are occasioned by technological developments in digital imaging. Technological developments, such as computer-wrought changes to the processes used to create images as well as the uses of images, are potentially more encompassing than those sparked by previous media technologies. While the introduction of the printing press altered distribution processes for communication, and the introduction of photography changed the composition of images, media studies scholar Lev Manovich argues, "the computer media revolution affects all stages of communication, including acquisition, manipulation, storage, and distribution; it also affects all types of media—texts, still images, moving images, sound, and spatial constructions" (19). The changes in what and how we communicate that Manovich enumerates affect the production of digital images in science no less than in other areas of our culture. While discussions of digital media (or "new media" [10]) tend not to include scientific images, the similar production processes and circulation of images made with the STM (as well as other digital scientific images) suggest that they could—and should—be discussed. *Haptic Visions* contributes to rhetoricians' developing body of knowledge about the rhetorics of digital imaging through analysis of the rhetorics of digital images in nanotechnology, especially through this book's focus on the effects of interaction on viewers, who are also users, and on the effects of communicating information in digital image form.

Digital media's inclusion of interaction is particularly interesting in relation to nanoscale images, as the focus on interaction connects in suggestive ways to Daston and Galison's identification of images-as-tools in instances of nanomanipulation. Manovich argues that digital images change our "concept of what an image is—because they turn a viewer into an active user" (183). The shift in image functions that Daston and Galison and Manovich (and others) observe raises questions for rhetoricians about the roles of researchers and image viewers in the production and reception of digital images that show nanomanipulation. Insights about what becomes persuasive in the context of nanomanipulation, and how those involved are persuaded to shift from viewers to users, may then also inform how other digital media operate. In addition, the exploration of the rhetorical effects of digital images on the scientific cultures from which the images emerge in *Haptic Visions* may also reveal insights about the development of digital media.

Studies of scientific digital images also contribute important analyses of how images shape and express information, especially because the primary function of images in science is to convey data. (However, the functions of scientific digital images are not limited to expressing information, as is discussed in Chapter 3.) Studying in detail how informational images—images designed to convey data[11]–function rhetorically in discourse is especially important because informational images do not create an inherent relation between the image's appearance or resemblance and what the image represents. For example, while the "IBM" images contain elements that "look like" atoms, the image elements are not actual depictions, like photographs; yet, the images are depictions—of *data* about objects built through interaction with the image and atoms. Digital, informational images blur the boundaries between presenting information and resembling objects—objects that cannot be visualized with the eye, or even with light microscopes, as atoms are smaller than the minimum resolution of light waves. Exploring how informational images such as STM images function rhetorically within discourse, given the blurred boundary between presenting information and resembling objects, can articulate how informational images function persuasively as well as how the images' form and expression of information affect the arguments in which the images appear. Studying the rhetorical role of images important to the discourse of nanotechnology can thus show the impact of digital media on a field, as well as a field's impact on digital media.

Toward A Visual Rhetoric of Digital Images in Nanotechnology

To explore the rhetorics of images in nanotechnology that exist at the intersection of emerging science and technology and digital data visualization, such as those Eigler and Schweizer created, *Haptic Visions* considers how instruments, scientific imaging practices, and image-viewing practices may influence and produce rhetorics. To do so, this book focuses on how, in addition to producing images, the particular configurations of practices that allow the STM to visualize and manipulate atoms also make for a new visibility. Atoms do not only become "visible" in the general sense of the word, showing up on the computer screen or in journal articles, but also more specifically become places within what Michel Foucault calls a field of visibility. As Gilles Deleuze explains in his book on Foucault, fields of visibilities are not only physical spaces, but also are ways of distributing light: "If different examples of architecture, for example, are visibilities, places of visibilities, this is because they are not just figures of stone, assemblages of things and combinations of qualities, but first and foremost forms of light that distribute light and dark, opaque and transparent, seen and non-seen, etc." (*Foucault* 57). Thinking of what becomes visible in terms of a field of visibility expands inquiry from what is or is not seen to how it becomes possible to see what is seen—how forms of light that distribute light and dark, seen and non-seen, for example, make it possible to see interaction with atoms.

Following the practices that allow interaction with atoms to become visible requires following more than the rhetorics of published STM images such as Eigler and Schweizer's; further insight into what and how STM images communicate also requires attention to the material and cultural conditions from which scientific knowledge about the nanoscale emerges. Richard Doyle's coinage of the term "rhetorical software" to describe the interactions of rhetoric (in verbal or visual statements) with hardware (e.g., instruments, lab equipment, etc.) and wetware (i.e., human users) articulates how material and cultural conditions constrain and enable the others in the production of scientific discourse (7). The concept of "rhetorical software" also highlights how digital images can persuasively communicate about the nanoscale through rhetorical registers that may be new, as well as through those that may be familiar, as rhetorics interact with hardware and wetware. *Haptic Visions'* study of scientific images' rhetorical functions within the context of STM images' production and use, as well as the field in which the images circulate, thus provides a detailed example of digital image rhetorics that expands the visual rhetoric of digital images beyond that which has been developed in conjunction with art, in new/digital media studies, or computer-mediated composition. This book also argues for the importance of material, embodied rhetorics of image production processes through focusing on the material and cultural conditions of image production and viewing, thus also contributing to scholarship of visual rhetoric and the rhetoric of science on production practices, and to scholarship on the embodied, material rhetorics of technologies. *Haptic Visions* extends analysis of the visual beyond representational analysis through studying images whose visual format does not always relate to what the images depict, so that analysis moves beyond how what an image *looks* like communicates its message. This book adds to recent work in visual rhetoric, such as the work of Bradford Vivian, who argues that we consider aspects of images besides representational ones as rhetorical. Vivian examines how images "*produce* and *enact* modes of spectatorship, subject-object relations, forms of affect, or grounds for competing attributions of sense and value in ways that cannot be explained in full by representational categories" (474). Though Vivian analyzes political images, his focus on the rhetorical aspects of images that reach beyond the representational, such as their productive capacities, is a focus that *Haptic Visions* shares. Following the operations of rhetorical software within the space in which the manipulation of atoms and interactions with the nanoscale become visible thus illuminates not only atoms, images, and the "software"—the rhetorics—but also the ways in which we are configured within that space of visibility.

The rhetorical dynamics surrounding STM images show how new visualization technologies and images become incorporated persuasively and understandably in communications. Studying the rhetorics of STM images also encompasses the study of how we are altered by the habit of interacting with STM images—we incorporate them into "our very humanity," as scholars like Donna Haraway (183–201) and Langdon Winner (12) argue. How do digital visualization technologies extend our notion of seeing, and how do we change in order to see what we cannot see with our biological eyes? How do such changes affect information and objects of scientific study? These broader questions form the impetus for the more specific analysis of STM images in discourses of nanotechnology that *Haptic Visions* undertakes.

Organization of Haptic Visions

To articulate a space in which atoms and manipulations are made visible, this book presents a close study of nanoscale images such as Eigler and Schweizer's to explore the rhetorical possibilities of STM images. At the same time, *Haptic Visions* traces the visual cultures from which the images emerge and in which the images circulate; this book also articulates the transformations occurring within science, knowledge production, and ourselves that allow us to see and manipulate atoms. I draw from work in rhetoric; the rhetoric of science, medicine, and technology; visual studies (particularly the history and sociology of scientific images, art history, and digital media studies); cultural studies; and science studies. I also draw from primary texts in physics, chemistry, and nanotechnology; and from interviews I conducted in 2005 and 2006 with scientists who use the scanning tunneling microscope.

To focus on manipulation, each chapter in this book analyzes a different aspect of how interaction functions rhetorically in the case of nanotechnology—in the workings of the STM, in Chapter 1; in changing vision practices associated with the use of the microscope and its productions, in Chapter 2; in changing visual conventions used in STM images, in Chapter 3; and in common but changing scientific tropes that appear in both text and image in nanotechnology discourses, in Chapter 4. While each chapter can be read independently, the different topics of the chapters overlap at times; together, the chapters articulate complex dynamics of the rhetorics at play in nanotechnology discourses and scientific visualization discourses.

Chapter 1, "Imaging Atoms, Imagining Information: Rhetorical Dynamics of the Scanning Tunneling Microscope," explores the question of how atoms become both visible and manipulable through focusing on the rhetorics of the visualization technology that Eigler and Schweizer used, the STM. In the chapter, I analyze the rhetorical possibilities inherent in the STM's operating dynamics, and then situate these operating dynamics within wider scientific, medical, and digital visualization trends in order to highlight how the STM's operating dynamics contribute to rhetorical actions and productions. I also follow the influence of the STM's operating dynamics on the productions of the STM—images information, and atoms—in order to show how the rhetorics that the operating dynamics make possible help present atoms as able to be manipulated and as tangible, individual entities. I argue for my method of analyzing image production practices, including the operations of visualization technologies alongside the usual object of rhetorical analysis: productions such as STM images.

Chapter 2, "Camera Haptica: Blindness, Histories, and Productions of Haptic Vision," examines another set of practices associated with generating STM images: the vision practices that STM users and other viewers engage in while viewing STM images like the "IBM" series. After discussing how seeing can be understood as a practice that is itself derived from culturally and historically specific practices, I argue that the practices the STM user and the STM image viewer employ should be considered practices of haptic vision, practices that fuse the senses of vision and touch. Haptic vision practices found in STM use are themselves partially formed by other practices of seeing; including practices associated with microscopy, informational imaging, and digital vision. Haptic vision practices also affect image viewers through the constitution of a different and dynamic relation between the observer and the observed than dominant vision practices that involve perspective. As I demonstrate in this chapter, these haptic vision practices thus also affect the composition and rhetorical impact of STM images.

Chapter 3, "Haptical Consistency: Emerging Conventions of the STM Image-Interface," builds on the first two chapters as it explores some of the rhetorical effects on images brought about by engaging in the interactive, haptic practices inherent in STM use, image creation, and image viewing. I focus on how the use of STM images as interfaces affects visual conventions of STM images, thus accounting for the ways interaction affects image production and reading practices. While STM imaging conventions include references to the dominant convention of linear perspective, STM imaging conventions also exceed dominant conventions, revealing a disjunction between how images function to communicate information and the current cultural conventions for reading informational, scientific images. I argue that other visual conventions are developing, and analyze what the emerging conventions reveal about the persuasive elements of STM images.

This chapter also presents a framework for understanding how the reading practices of digital, informational images and their attendant conventions create effects in readers that should be considered a crucial component of how digital, informational images function rhetorically in discourses.

Chapter 4, "Visual Intelligence: Reading the Rhetorical Work of STM Imagesin Tropes," shifts my level of analysis of the rhetorics of STM images to specific tropes. I follow two established scientific tropes that occur frequently in scientific and nanotechnology discourses: the trope of writing that Eigler and Schweizer's "IBM" images exhibit, and the conventional microscope trope of tiny worlds made visible. I demonstrate how the expression of these two tropes mutates in STM images, linking the change in the tropes to the image production and viewing practices that microscope users and image viewers undergo. The alteration of common scientific tropes suggests changes in the formations of scientific knowledge and the field of nanotechnology, affecting our positions in relation to the nanoscale.

Finally, the conclusion, "Haptic Visions of Science and Rhetoric: Interaction and its Implications," explores the rhetorical elements identified in the previous chapters for how the elements may suggest possible changes in rhetorical strategies, given the deeply interactive nature of image-making, image-viewing, and the practices of haptic vision. In addition to signaling a change in rhetorical practices to include more persuasion through the experience of the image as interface, I explore why understanding more experiential, visual persuasion is important for understanding nanotechnology and other emerging sciences, especially in relation to the production of scientific knowledge. Such considerations become particularly important as practices of envisioning and arguing shift as we respond to changes such as those brought about by developments in digital media and visualization technologies, in addition to the ensuing changes in how we view, envision, and interact with the world from atoms on up.

1 Imaging Atoms, Imagining Information: Rhetorical Dynamics of the Scanning Tunneling Microscope

A curious thing happened to scientific concepts of atoms around the time that D. M. Eigler and E. K. Schweizer published the "IBM" images. Discussions of atoms drifted from the dominant quantum-mechanical perspective in which individual atoms cannot be measured because, as quantum physicist Erwin Schrödinger explains, "The individual particle is not a well-defined permanent entity of detectable identity or sameness" (qtd. in Regis 155). According to this view, atoms do not occupy one place at one time, do not have boundaries, and cannot be measured singly: atoms can only be measured in collective quantities. After the 1980s, however, scientists studying atoms returned to talking about atoms as individual, bounded entities—similar to how Isaac Newton envisioned atoms, although with a twist: now atoms were manipulable, almost tangible. The shift towards conceptualizing atoms as manipulable has important consequences for how we understand and move within the world around us. The shift towards the manipulable atom also forms a rich site for examining rhetorical practices in technologies, scientific fields, and the cultures in which the technologies and fields exist.

Why the shift towards the manipulable atom happened is not entirely clear. In a popular history about the early development of nanotechnology, Ed Regis contends that the events that helped spur what he calls a paradigm shift in the concept of atoms in the early 1990s include Richard Feynman's 1960 speech "There's Plenty of Room at the Bottom" and Robert Van Dyck, Philip Ekstrom, and Hans Dehmelt's capture of individual electrons in 1976 (Van Dyck, Ekstrom and Dehmelt 776).

However, Regis and others also suggest that the scanning tunneling microscope (STM), and related scanning probe microscopes, played a part. Science studies scholar Jochen Hennig, for example, argues that the STM has occasioned a shift in the definition of atoms ("Images"). Regis claims that Eigler and Schweizer's "IBM" atom manipulation proved that atoms "were in fact and could be treated as mechanical, Newtonian entities. They were objects that could be made to *do* things" (269).

Scientists also point to the development of the STM and related microscopies when accounting for the shift toward understanding atoms as manipulable, individual entities. For example, one scientist I interviewed in 2005 explained the conceptual change in terms of the development of techniques:

> When people first started thinking about surfaces and working on them scientifically, they thought in very atomistic ways, but they didn't really have techniques to look at them. They just kind of came up with ideas. And then as techniques developed, the techniques tended to be what was called reciprocal space [as opposed to "real space"], so they're diffraction measurements and people started thinking about periodic structures and then trying to guess what those structures were and compare them to their data. And then when scanning probes [that collect real-space data] came along, two things happened. [One was that] [t] he view flipped back to the atomistic view. . . . [12]

In another example from the journal *Science*, two scientists who use the STM, James K. Gimzewski and Christian Joachim, summarize the impact of STMs on scientists' relations to atoms: "By the early 1980s, scanning tunneling microscopy (STM) . . . radically changed the ways we interacted with and even regarded single atoms and molecules" (1683). Gimzewski and Joachim's implication that scientists expect to interact with atoms—so much that interacting with atoms is ranked as more important than regarding atoms—indicates the importance of interaction. In Gimzewski and Joachim's view, supported also by Eigler and Schweizer's images among other instances, atoms are not simply solid masses; atoms are also masses with which humans can interact.

The assertion that humans can interact with *individual* atoms affects scientists working with the STM as well as the development of nanotechnology, as mentioned in the introduction. While rhetorics of arguments that use atom manipulation as justification for developing and funding nanotechnology reveal fascinating dynamics of policy and field formation, the rhetoric of atom manipulation through interaction with the nanoscale also functions at the level of everyday scientific practice, including imaging. The fact that discourses about atoms came to include assumptions of interaction for scientists also prompts further questions for rhetoric. Statements from the scientists and historians mentioned above, for example, suggest the STM is a key player in this change, although the STM did not create the first views of atoms—Erwin Müller first photographed atoms in 1955 with a field ion microscope (Müller and Bahadur; Müller). Whether or not the STM affected the shift towards understanding atoms as manipulable, the fact that the STM is mentioned in discourses about this shift suggests that the STM exerts some influence on scientific practice and discourse.

This chapter explores what is persuasive about the STM and the rhetorics of the STM's operating dynamics in visualizing the nanoscale, contributing to a partial[13] account of how it is that atoms become both visible and manipulable. I identify rhetorical possibilities the STM includes and encourages through identifying and analyzing STM operating dynamics in the context of the STM and broader scientific, medical, and digital visualization trends. I also track the influence of STM dynamics on images, information, and atoms to demonstrate rhetorical links between the operating dynamics and productions of the STM. In so doing, I present (and argue for) a method for studying the rhetorics of visualization technologies that includes analysis of productions and production practices. Insights developed from analyzing STM dynamics individually, and in connection with STM productions, can further inform analyses of STM images as well as the scientific and other discourses to which the images contribute, including analyses found in this book. Before turning to the STM's dynamics, I demonstrate how scientific and medical instruments make rhetorical contributions to scientific and medical discourses, and articulate the STM's relationship to other recent scientific visualization technologies to show the significance of the STM for rhetoricians of science, medicine, and digital technologies.

Visualization Technologies as Rhetorical Instruments

Instruments contain much more than springs, circuits, and film, and they are certainly not (pace Bachelard) merely reified theories. Instruments embody—literally—powerful currents emanating from cultures far beyond the shores of a master equation or an ontological hypothesis.

—Peter Galison

Optical devices constitute "points of intersection where philosophical, scientific, and aesthetic discourses overlap with mechanical techniques, institutional requirements and socioeconomic forces."

— Jonathan Crary

Historian of science Peter Galison and art historian Jonathan Crary's comments on scientific and optical instruments exemplify some of the ways in which scholars from science studies, history of science and technology, art history, and related fields analyze how instruments develop and operate within complex systems of social and technical knowledge production. However, not many scholars have focused on the rhetorical aspects of instruments within knowledge-producing systems. Instead, most rhetoricians focus on either the productions of scientific or medical instruments (such as specific images or texts associated with knowledge produced by instruments) or the practices afforded by tools or technologies of instruments more directly related to composition (such as the computer).[14] However, elaborating on how scholars have studied the functions of such instruments offers support for an argument about how scientific instruments, such as the STM, include rhetorical functions. While researchers using new instruments can use an appeal to novelty to argue for the significance of the instrument, rhetorical appeals based on newness are limited in time and scope. Instruments can function in more potent rhetorical ways and influence the generation and practice of rhetorics that affect discourses surrounding knowledge production.

Instruments as Rhetorical Entities

Scholarship in the history of science, science studies, and art history features arguments for considering instruments as situated within the complexities of practice. For example, scholars within laboratory studies, a subfield of science studies, have generated ethnographies of the work of scientists that situate instruments within everyday practices, and describe how instruments function within complex systems of knowledge production.[15] While not focused on rhetoric, studies that situate instruments within the context of practice are also useful for understanding the rhetorical contributions of instruments within the context of scientific practice. Two ways that historians, science studies scholars, and art historians situate and analyze instruments also demonstrate the significance of analyzing the rhetorical aspects of instruments.

First, many scholars of science studies and related fields have used Bruno Latour and Steve Woolgar's theorization of laboratory work in their classic ethnography, *Laboratory Life*, to structure analyses that account for the wider cultures that Galison and Crary mention, as well as the daily complexities of scientific practice. Latour and Woolgar conceptualize the goal of laboratory work as the production of inscriptions—textual or visual marks on paper in scientific journals. The practices that produce inscriptions are the work of the laboratory; inscriptions form part of how scientists convince other scientists that statements produced by scientists are scientific fact, are worth passing along, and are worth citing (Latour, "Drawing" 24). Indeed, Latour claims that inscription practices lead towards rhetorical as well as epistemic goals: Significant uses of writing or visualization in science are "those aspects that help in the mustering, the presentation, the increase, the effective alignment, or ensuring the fidelity of new allies" ("Drawing" 24).

Focusing on inscriptions while analyzing scientific practice highlights the rhetorical possibilities of instruments, especially as inscription practices can include devices (such as instruments) as well as communicative methods: material and social practices inform and contribute to scientific knowledge production, and also influence scientific method.[16] Some rhetoricians also acknowledge the usefulness of Latour and Woolgar's concept of inscriptions. Jeanne Fahnestock, for example, calls for rhetoricians to "come to terms with the many techniques of visual inscription used to generate evidence" in order to develop a visual rhetoric of science ("Rhetoric" 284). In a recent article, Chad Wickman uses the concept of inscription to analyze technical laboratory practices and text production, focusing on how visual representations become rhetorical objects in scientific practice ("Observing" 152). While Wickman focuses on visual representations, Latour and Woolgar's concept could also be used to analyze technical practices used in the laboratory and in image production, with attention to how instruments become part of the rhetorical process inherent in inscription practices.

Second, as part of their function as inscription devices in the laboratory, the operation of instruments requires that users engage in practices specific to the instruments. Over time, instrument users become habituated to the practices necessary to use instruments.[17] The fact that users do become habituated experientially to certain practices through instrument use is another reason to look carefully at instruments as rhetorical entities: Instruments help establish repeated, everyday practices of knowledge-making, or may employ practices already used by a community. Media studies scholar Scott Curtis presents a fascinating example of the influence of a community's habits of observation in the introduction of photography as an observational technique in medicine. Curtis argues that the adoption of photography depended not only on its automation, but also on its presentation of "a set of features that spoke to established and emerging principles and habits of observation" (85). Habituated practices of using instruments may influence users in ways that subsequently affect the development or use of an instrument, or that affect the expression or formation of the inscriptions an instrument produces. Thus, how specific practices become habituated, and how practices affect the uses of other instruments or other practices, becomes relevant to understanding how science works. Understanding how habituated practices of instrument use inform rhetorical practices then becomes important in order to fully account for the rhetorics of visual inscriptions. (Chapter 2 discusses the establishment and habituation of one set of vision practices.)

Attention to habituated practices of instrument use also contributes to the rhetoric of scientific processes and practices. Wickman and Heather Brodie Graves argue for considering scientific practices and processes as analytic objects of study for rhetoric, including the rhetoric of the processes of scientific inquiry (Wickman, "Rhetoric;" Graves, 81-142). Graves, for example, analyzes the language physicists use when engaged in the process of making science; Wickman includes nonlinguistic modes in his analysis, discussing and extending Aristotle's concept of *techne* to argue that processes of scientific inquiry are rhetorical. Doing so creates a space for the rhetorical analysis of scientific practice amidst changes to that scientific practice that are spurred by changing technologies (23), including a focus on how "visuals and other nonlinguistic modes contribute to scientific meaning-making and persuasion" (23). Arguments from Wickman and Graves sketch some ways in which scientific practices and processes are rhetorical; a focus on instruments such as the STM extends the argument for analyzing processes and practices by exploring how instruments used in everyday practice and production create rhetorical possibilities.

The STM as Visualization Technology: Context, Characteristics, and Rhetorical Significance

Instruments are not only important because of their everyday use in the laboratory. Instruments are also important because they are situated within broader, complex trends in scientific and technical development. Situating an instrument within broader trends can illuminate cultural influences that are common to more than one instrument; therefore, analysis of one instrument may provide insights for the study of other instruments. The STM shares some rhetorically significant characteristics with other recent scientific and medical visualization technologies, such as PET scans, CAT scans, MRI, fMRI, and ultrasound. These rhetorically significant characteristics include the presence of complex mediation and interpretation practices, as well as the use of images to arrange and deliver large amounts of data.

A brief summary of how the STM works illustrates the complexity that Galison and Crary discuss, and introduces characteristics common to recent digital visualization technologies. To use the STM, researchers apply a weak current to the conductive metal tip of the STM. The tip is then brought close to the surface of a conductive substance (i.e., the sample). Although a gap exists between the tip and the surface atoms, when an electric charge is applied to the tip, the electrons of the surface atoms "tunnel" through the space (frequently a vacuum) between the tip and the surface to interact with the atoms on the tip (hence, the "tunneling" in the microscope's name). The tunneling, a quantum effect, thus changes the voltage of the tip because the tunneling electrons, like all electrons, are charged. The electron cloud is exponentially sparser the further the electrons travel from the atom's nucleus, and the voltage changes according to the density of the cloud, so that the voltage change is consistent with the distance from the surface atoms. The tip passes just above the surface of the sample using a piezoelectric element (or piezo) that either slightly expands or contracts when voltage is applied, and that produces an electric current when pushed, enabling fine control and measurement of voltage changes produced by the different densities of electrons at each spot. The STM operates in one of two modes: a constant-current mode, which keeps the current of the tip constant by allowing the tip to move up and down, depending on the electron densities the tip encounters from the surface; or a mode that measures the current of a tip that does not move up and down, but instead keeps a fixed distance from the sample. The STM then collects measurements of changes in either the movement of the tip or of the current at fixed intervals as the tip scans the surface.

To create an image from these measurements, the STM displays the data on a computer monitor, arranging the data in a matrix in the order of the sampled measurements and assigning each a value. Values are then sent to a computer monitor that assigns each value to a pixel.[18] The measurements are expressed in pixels as different values in a gray scale or false color scale ("false" because atoms do not have colors: light waves are too large to optically register atoms) to create more visible variation; the accumulated matrix of values, presented in pixels, comprises the image. Pixels can be arranged in numerous formations to construct and communicate data in spatial arrangements (what we think of as a digital image, generally) or as histograms that graph numerical frequencies of the values assigned to the pixels. The imaging of data by the STM highlights the differences between measurements, and so makes visible the topographic or electronic properties of the sample.

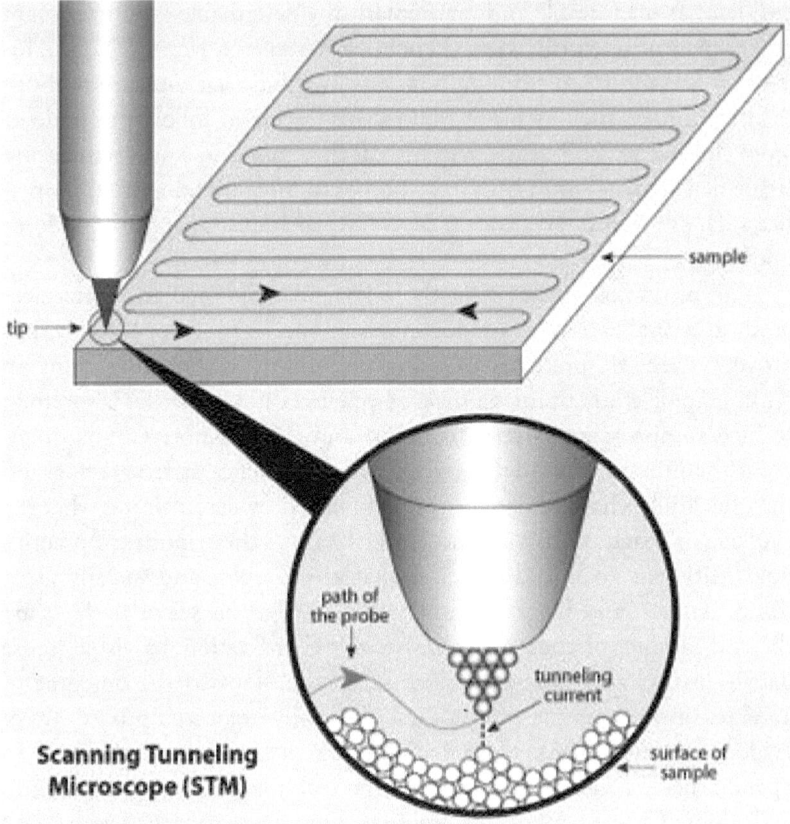

Figure 3. "How an STM Works." David Beck, (c) Exploratorium, www.exploratorium.edu.

Mediation and Interpretation Practices

As this summary of the operation of the STM suggests, the instrument uses extensive processes to mediate between data and image in converting measurements of phenomena into data, and then into images. Many scientific and medical visualization technologies that produce digital images engage in similarly complex processes of mediation to visualize the invisible—whether invisibility is due to location, such as within a living body (as ultrasound, CAT, PET, and fMRI technologies visualize); or to size, such as below the threshold at which light operates (as probe microscopes such as the STM visualize); or some other invisibility.[19] Digital scientific visualization technologies frame phenomena (such as tunneling data) as measurable and mathematically describable—an important component of scientific work (Lynch, "Externalized Retina" 170). This is often accomplished through non-lens methods that measure non-optical attributes, such as using radioactive tracers to follow molecule or atom flows (e.g., PET scans, MRI, fMRI), ultrasound waves to measure tissue density (e.g., ultrasound), or probes to measure non-visual properties such as atom interactions (e.g., STM) or friction (e.g., atomic force microscope).

The mediation practices of the STM, like the mediation practices used in visualization technologies mentioned above, operate within broader scientific practices of using visualizations. Drawing from an ethnographic study of image-making practices in a scientific laboratory, science studies scholars Klaus Amann and Karin Knorr Cetina document the process of producing visual data in science as first seeing and then deciding what the data is; this is followed by determining what the evidence is. Each of Amann and Knorr Cetina's three modes of practice holds different goals and practices that are complex and socially organized. Amann and Knorr Cetina comment, "just as scientific facts are the end product of complex processes of belief fixation, so visual 'sense data'—just what it is scientists see when they look at the outcome of an experiment—are the end product of socially organized procedures of evidence fixation" (86). The mediation practices that researchers use to operate the STM and other visualization technologies fit within Amann and Knorr Cetina's modes of practice. The practices that Amann and Knorr Cetina identify include multiple processes of mediation that are not only based on scientific practice, but may include broader social practices, thus indicating, one way in which visualization technologies form points of intersection for the "mechanical techniques, instrumental requirements, and socioeconomic forces" that Crary mentions (8). Therefore, the practices of mediating between data and image also form a site for analyzing the rhetorical work of instruments.

Along with extensive mediation practices, inscriptions tend to require extensive interpretive practices to understand what is being shown; although, visual inscriptions may present phenomena in ways that look simple or apparent. Amann and Knorr Cetina's study underscores the complexity of the processes of visualization and interpretation in making scientific knowledge. The apparent simplicity of the image as a form for presenting data sometimes leads to confusion about how to read visual inscriptions made by instruments, even by experts.[20] For example, in the early days of the STM, researchers occasionally misinterpreted what the images showed (Mody, *Instrumental* 12–13; Woodruff 75). In a study of PET scans, Joseph Dumit explains that some of the difficulties researchers have in interpreting scans involve not only habits of seeing, but also theoretical questions of interpretation (68–69).[21] Thus, practices of mediation and interpretation of the productions of instruments form two aspects of the complexity of digital visualization technologies used to create inscriptions; interpretive practices, like mediation practices, suggest possible directions for analyzing rhetorical functions.

Images of Data

Galison's account of the merging of two traditions of presenting evidence in microphysics in the early 1970s points to an origin for the development of recent scientific and medical visualization technologies that present large amounts of data in image form, such as the STM, MRI, and PET scans, for example (Image and Logic 570). Galison argues that the development and use of electronic images in physics for the first time allowed researchers from these two traditions to combine methods and see images as evidence. Because large amounts of data could compose images, researchers who had focused on images like photographs for evidence could use the same methods—images—as those researchers who relied on statistical data for evidence. The use of images to present evidence authoritatively within the field was so important, Galison argues, that "the controllable image" has become the main form that data has taken in the sciences (Image and Logic 810).[22] Indeed, many scientific and medical visualization technologies that rely on data in image form were developed and used in the 1970s.[23] The STM, too, was developed within the trend of using data to compose images: The inventors of the STM first filed for patents in 1978 in Switzerland, and then in 1980 in the US (Granek and Hon 102).[24]

The expression of data in image form, common to visualization technologies such as the STM and the MRI, raises questions about how the use of data-images, or informational images, affects the ways in which images function in discourse and in reading practices or assumptions about what the images convey. Amann and Knorr Cetina's observation that the significance of data must be discovered highlights the importance of exploring how the presentation of data in electronic and image form affects the practices scientists use to discover or interpret the data. The fact that concerns about communication with informational images is common to a group of digital visualization technologies such as PET, MRI, STM, and ultrasound suggests that while the details of the context of practice for each visualization technology offer unique insights, a study charting the rhetorical effects of one visualization technology (like the STM) can be relevant to the study of the others. Further, few studies exist on the rhetorics of the creation of data images by these technologies, or on the rhetorics of digital images such as those the STM, MRI, and related technologies produce.[25]

The complex mediation and interpretation practices needed to express data in images also affect users in ways that suggest directions for further study of the rhetorics of instruments. Scholars in art history, science studies, and media studies recognize the importance of investigating the effects of an instrument on the experimenters or users of the instruments; the insights of these scholars provide a basis for extending analysis to rhetorical aspects. For example, Crary argues that optical instruments reconfigured the position of the nineteenth century observer, creating a change in visual practices (see 8-9 for a brief summary of his main argument as it relates to instruments). Galison and Daston examine the identity of the scientist in *Objectivity*, recounting how the scientific self has been shaped by historically specific scientific practices that are associated with the main "epistemic virtue" of the time, such as objectivity (191-251) or, more recently, trained judgment (357-61). Curtis explores the ways in which photography trains medical observers. While Crary, Daston and Galison, and Curtis do not focus on rhetoric, their detailed analyses of instruments in practice present historical and social models of how instruments may affect users, perhaps informing rhetorical analyses of how users are influenced by using instruments.

A Method for Analyzing Material, Embodied Interactions

This chapter performs a close reading of the operations and productions of the STM as it demonstrates how the specific rhetorical influences of a technology affect its productions (such as STM images). I focus on the rhetorics of the STM as part of the complex, material, and embodied practices of scientific knowledge-making and communication. Attention to the production processes of technologies is likely to become more and more important for analyzing the productions of those technologies, whether those productions take the form of images, databases, or other digital objects; how we make arguments with visualization technologies can be informed by the productive capacities and effects of those visualization technologies. In performing a close reading of the operation of an instrument, and considering the ways in which the STM is imbricated within social and material practices as an inscription device, this chapter highlights the material and embodied rhetorics at play in the formation of scientific knowledge, adding a rhetorical dimension to science studies of embodiment, and a science studies dimension to rhetorical studies of embodiment.[26]

My close reading of the STM through the dynamics of its operation also necessitates analysis of interaction; interaction is crucial to the operation of the STM. As the summary of the STM's operations above suggests, creating images of the nanoscale involves interactions between electrons, material apparatuses (including scanning and computer components), human actions, computer software, and cultural practices—practices of seeing and organizing as well as coordinating information and things. STM operation relies on interactions among the constitutive elements of the instrument, interactions that exist not only between user and sample, as in most microscopical work (Keller, "The Biological Gaze" 112), but also at other levels, such as the interaction between the sample and the STM. Therefore, closely reading the interactions that compose STM dynamics reveals rhetorical dimensions of scientific practices of making knowledge.[27]

What interaction means, however, also becomes a question: In casual use, "interaction" is often over-generally applied to anything involving a computer, and the term's definition is disputed among scholars studying digital media and technology.[28] Janet Murray clarifies that what is often meant by interactivity in computers is that "they create an environment that is both procedural and participatory" (74). Rhetoricians analyzing the context of writing in new media or digital media environments focus on the participatory aspect of interactivity, without paying much attention to the participatory aspect that Murray identifies, using the term to indicate the ability of the user/reader to communicate with the user/writer. A few rhetoricians, such as Teena A. M. Carnegie, James Porter, and Ian Bogost, however, examine procedural as well as participatory aspects of interaction.

Porter, Carnegie, and Bogost suggest that the value of interaction lies in shaping audience response as a structural element, as opposed to only content—that form and content are both components of the rhetorics of interaction. Carnegie argues that the computer interface functions as a Ciceronian exordium, a rhetorical opening strategy that aims to engage the audience so that the audience is receptive to hearing the argument (165). Carnegie claims that three modes of interactivity, drawn from new media and human-computer interaction (HCI) research, multi-directionality, manipulability, and presence, are also the rhetorical modes of the interface (166). Carnegie links these three modes to higher levels of audience engagement, drawing from Sheizaf Rafaeli and Fay Sudweeks's finding that encouraging interactivity in users produces higher levels of engagement (166). Carnegie's exploration of how the rhetorics of engagement function in visuals such as interfaces helps make interfaces visible as rhetorical sites that structure interaction. Carnegie's focus on modes of participatory interaction further articulates details of how the user becomes involved in the operating dynamics of an instrument.

Porter's description of digital composition includes interaction as one element that writers in electronic spaces need to consider in their delivery decisions, acknowledging that "different types of computer interfaces and spaces enable different forms of engagement" ("Recovering" 217). Porter suggests that interaction forms a significant part of productive rhetoric for writers in digital spaces, especially with attention paid to how participation and procedure may be linked. Like Carnegie and Porter, Bogost develops a structural analysis of the rhetorical components of interaction, emphasizing the procedural rhetoric of interactivity in the context of videogames. Bogost argues that interactivity can be understood in relation to the Aristotelian enthymeme, in that videogame players supply the warrant as they play (43). Interaction allows the player to proceed through the content, or "argument," of the game in ways that are mental, but also are embodied—such as using the joystick or pressing keystrokes to respond. Bogost also analyzes the procedures of computer games to interpret messages created by how computer games structure narratives according to user responses. While Bogost does not analyze the production processes of procedural rhetorics per se, his approach of considering the structuring of interactivity as rhetorical—and involving the user as well as elements of what the user engages with—is one applicable to production practices such as those used to create interfaces.

The concept of *affordances* presents one way to focus on what interactions may be most significant, and as such, most significant for understanding the rhetorics of procedural and participatory components of interaction. Carolyn Miller argues for understanding parallels between rhetoric and technology (particularly communication technologies) through the idea of affordances, quoting psychologist James Gibson's formulation of affordances as what an environment "provides or furnishes, either for good or ill" (Gibson 127). In Miller's view, affordances help explain how technology, like rhetoric, can lead users toward some possibilities, and away from others. Miller argues for how affordances of communication technologies function:

> [A]ffordances take the form not of material properties or ecological niches [as they do for physical environments like an animal's habitat] but rather properties of information and interaction that can be put to particular cognitive and communicative uses. Thus a technological affordance, or a suite of affordances, is *directional*, it *appeals* to us, by making some forms of communicative interaction possible or easy and others difficult or impossible, by leading us to engage in or to attempt certain kinds of rhetorical actions rather than others. (x)

Following how a technology's affordances create particular responses in users reveals rhetorical possibilities that the affordances encourage, and even create. Extending Miller's direction, following affordances helps articulate what properties of information and interaction encourage rhetorical actions in operating a particular technology as well as what kinds of rhetorical actions are most encouraged. Tracing the persuasive in the affordances a technology creates thus provides a way of exploring interaction while attending to an instrument's productive, material rhetorics. Following the affordances a visualization technology creates, then—and going back to Michel Foucault's architecture of making visible, as discussed in the introduction—helps to identify some of the available possibilities that shape what can become visible.

As I analyze the operating dynamics of the STM in light of their affordances, I also observe the interactions the operating dynamics encourage in relation to four main ways of characterizing interaction from HCI (human-computer interaction) studies in order to explore how the interaction configures possibilities for the user. In a survey of different views of interactivity from the HCI research traditions, Sally McMillan explains that the following four ways include three that are based on Claude Shannon's model of communication as information a sender communicates to receiver: user communicating to computer; computer communicating to user; and an equal, adaptive interaction where "the computer is still in command of the interaction, but that it is more responsive to individual needs" (McMillan 175). The fourth constitutes interaction differently, in terms of what Mihaly Csikszentmihalyi refers to as "flow:" it "represents the user's perception of the interaction with the medium as playful and exploratory" (qtd. in McMillan 173–4). "Flow" tends to include participation from both sides, so that neither computer nor user occupies either "sender" or "receiver" roles; instead, computer and user take on both roles, and so become co-creators or participants (McMillan 174). As McMillan further describes, flow is

> characterized by a state of high user activity in which the computer becomes virtually transparent as individuals 'lose themselves' in the computer environment. Virtual reality systems seek this level, but it may also be characteristic of gaming environments and other situations in which the user interfaces seamlessly with the computer. (175)

These four models of interaction present different experiences for users; the interaction models may also generate different patterns of user response, further determining how users interface with the computer through the screen. Comparing STM dynamics to the models of interaction provides a more specific sense of how STM dynamics are structured.

Studies in rhetoric, science, and HCI identify "interactivity" as the structuring of events that involve writers/users, media, knowledge and information, and users/readers in particular configurations. A focus on "interactivity," then, is also a focus on finding, describing, and analyzing interfaces—places of interaction, boundaries where forces or disparate elements meet. This chapter begins my focus on interactivity as I identify, describe, and analyze interfaces; Chapter 3 and Chapter 4 further analyze interfaces. In the next section, I analyze STM dynamics to identify the affordances that instruments create that, in turn, impact the shaping of inscription practices scientists engage in when using the STM. The STM dynamics influence the form of the inscriptions that help create scientific statements, and so shape nanotechnology and the concept of the atom. The affordances shape rhetorical possibilities inherent in the inscription practices that are also visible in the inscriptions themselves, as I explain below.

Manipulating Atoms: Microscope Interactions

Three main dynamics within extant visualization and instrumental traditions in science and related technologies help constitute the visualization practices of the STM: electron tunneling, raster scanning, and image processing using a graphic user interface (GUI). Each of these three dynamics structures interactions between apparatus, user, data, and the nanoscale; informs how instruments mediate the transformation of phenomena to data and to image; and helps structure how scientists interpret the data in the image. While each of these main dynamics functions separately to some extent, the interactions between the dynamics combine, expanding connections and enhancing the intensities that each may possess alone. The coordinated interactions of the dynamics of electron tunneling, raster scanning, and GUI image processing then enable the STM to function, producing and arranging data about the nanoscale, thus affecting what STM images convey and how the images do so. The coordinated interactions of the STM operating dynamics structure the possibilities for making atoms visible—and also help create the productive rhetorical possibilities of the STM.

Electron Movement: Tunneling Electrons and Interactive Surfaces

One of the major dynamics on which the design of the STM is based relies on the interactions between a conductive surface (composed of a metal, for example) and the microscope tip, as the tip does not contact the surface, but remains about a nanometer away (Mantooth 9). Instead of contact, the interaction between tip and surface is a result of electron tunneling. Tunneling is based on the articulation of electrons as both particles and waves in quantum mechanics, where "each electron behaves like a wave: its position is 'smeared out'" (Binnig and Rohrer, "The Scanning Tunneling Microscope" 52). The behavior of electrons as both particles and waves allows surface electrons to "tunnel" through the barrier of the vacuum between surface and tip atoms, and thus interact with the electron cloud of the atoms or atoms on the tip. Measurements of the tip's electron clouds through voltage thus presents a way to understand the surface atoms through the behavior of the behavior of the atoms. Use of electron tunneling as a measurement technique in the STM is part of a broader trend in creating images from non-optical data, and has implications for what is able to be visualized with the instrument.

Electron tunneling is a relatively new idea; the incorporation of electron tunneling into the STM shows how the dynamic fits into the larger story of the development of non-lens-based visualization technologies. In 1960, Ivar Giaever first published the results of demonstrated electron tunneling (Giaver 147–48). For his research, he received the Nobel Prize in 1973. However, scientists did not apply electron tunneling to instrument development until the early 1970s, when Russell Young, John Ward, and Fredric Scire created a machine called the "topografiner" that, like the STM, used electron tunneling and three-dimensional scanning to measure "the microtopography of metallic surfaces," but used a field emitter instead of a tip to create tunneling conditions (Young, Ward, and Scire 999). The topografiner was not very successful in achieving measurements due to interference from outside vibrations, caused by people walking in the building, for example. In the early 1980s, STM inventors Gerd Binnig and Heinrich Rohrer, along with Christoph Gerber and Edmund Weibel, reduced outside vibrations enough to measure the tunneling and develop the STM (Binnig et al. 178–180).[29] The story of how electron tunneling became a measurement technique illustrates the complex mediation required of some visualization technologies, mediation anchored in the practices of a larger community.

The use of electron tunneling to visualize atoms affects what can be measured as well as the relations between the different atoms interacting at the interface of the vacuum. What is measured is the atomic movement that enables electron tunneling, not an object such as an atom. Recording the interactions of electrons with other electrons in a vacuum also transforms the distance between tip and surface electrons into a dynamic interface. To create data points, then, the tip passes across the area to be imaged, sampling the changes in voltage produced by the different densities of electron interactions at different spots. Therefore, the STM maps encounters, local events of tunneling.

One effect of the use of electron tunneling to create measurements is that both tip and sample can affect the interaction (and thus the measurement). For example, the tip's characteristics can affect the sample and the image produced from the sample, establishing a multi-directional affordance that both creates and structures the dynamic between tip, sample, and resulting image. As STM textbook author Chunli Bai explains: "The size, nature and chemical identity of the tip influence not only the resolution and shape of a STM scan but also the electronic structure to be measured" (9). For example, the conical shapes that form the rim of the Quantum Corral (Figure 4) image are not "what atoms look like;" instead, the shapes are effects of the tip's V-shape, and reflect the tip's traverse from one level to another while, at the same time, moving across the sample surface. In short, the shape is more like a graph of the tip's movements over time (Russell). One result of the mutual influence of tip, sample, and image is that the shape of the tip becomes important: Ideally, one atom at the tip end should protrude slightly (even just an Ångstrom) from the others so that the applied current can flow through that one small point (J. Foster 17).[30] The sample, too, can affect the tip if the atoms of the sample strongly attract and "pull off" some of the tip's atoms. In the space of the interaction, both tip and sample meet in the vacuum, and are equally able to affect the interaction.

The use of electron tunneling also creates the possibility of continued interaction, like MRI or PET scanning, as neither sample nor tip is damaged in collecting measurements (unlike, for example, electron microscope samples that are destroyed in the imaging process). The fact that the sample is not destroyed allows researchers to collect data repeatedly over the same space, and thus track dynamics over time; the user can experience atoms as a series of movements. A *Journal of Physical Chemistry B* article provides an example: S. A. Kandel and P. S. Weiss state that "by comparing sequentially recorded images, [they] can see that the size and shape of the clusters [on the sample] change over time" (8103). Kandel and Weiss explain the effect of the tip on the measurement: "the mobility shown in the rearrangement of these clusters is likely (at least in part) induced by the STM tip" (8103). The dynamic that occurs between the tip and the sample as the STM collects data is similar to the HCI interaction type of "flow," where a user experiences a merging with the computer. In this case, of course, the interaction occurs between atoms. Although the tip-sample interaction does not directly involve the user, electron tunneling affects the user's experience of atomic phenomena through a multi-directional affordance and the related possibilities for repeated interaction. The structured dynamic of the tip-sample interaction that also produces nanoscale measurements affects what can be shown in an image and affects experiment design, just as the use of electron tunneling allows researchers to manipulate the surface using the microscope's probe affordances. Thus, the expectation of a certain kind of multidirectional interaction inheres at the level of collecting data.

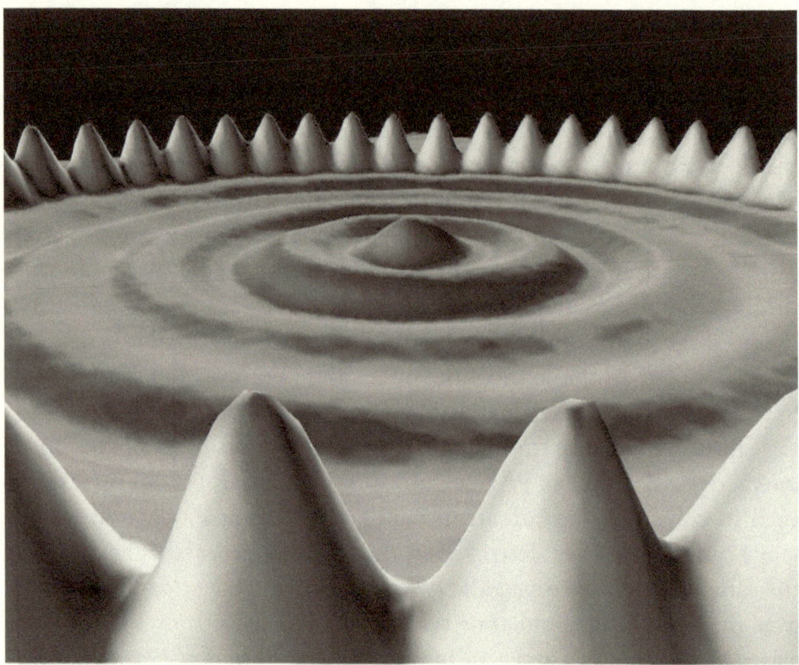

Figure 4. "Quantum Corral" image. From *Science* 262, 5131 (8 Oct., 1993). Cover illustration. Image originally created by IBM Corporation. Reprinted with permission from AAAS.

Raster Scanning and Z-direction Moves

The STM tip moves in three dimensions (x, y, and z, in the Cartesian system), forms other constitutive components, and works with electron tunneling to shape the multi-directional affordance of the STM. The microscope's design, like that of the scanning electron microscope, relies on a mobile tip. The fact that the tip moves—and must move—to survey the sample affects how the STM measures each point. The tip moves in three different directions: in the x and y directions in a raster pattern to collect data about points on the surface, and in the z (up and down) direction to create the tunneling conditions in which tip and surface electrons interact. Movements in x, y, and z directions coordinate with tip-surface tunneling interactions to collect data about the nanoscale; the x, y, and z movements also are part of broader traditions of dynamics used by other visualization technologies. In the STM, the x, y, and z directional dynamics structure interactions that encourage manipulation by the user.

The process of rastering, or scanning the surface with a back-and-forth pattern, from which the STM builds an image from data, forms a key dynamic in other visualization technologies. Derived from radar's creation of an image from a signal, rastering is perhaps best known as the method that allowed cathode ray tubes, such as those used in non-digital televisions, to build an image from a linear signal. In microscopy, using rastering to image a sample had been considered since the early twentieth century, but use of rastering was first demonstrated in 1972 (Wickramasinghe 78). The STM-like topografiner also scanned microscopic surfaces and collected measurements to create an image of the surface, although the specific pattern the scanner makes is not described (Ward, Young, and Scire 999). The incorporation of rastering into scientific visualization instruments such as the topografiner in the early 1970s also fits into the trend of developing visualization technologies that Galison discusses. The STM, then, is partly formed by the "tradition" of dynamics of rastering.

In the STM, rastering only works in conjunction with measurement of tip-surface interactions in electron tunneling, because the dynamics involved in raster scanning are structured around the challenge of movement in relation to time. To create anything other than a blur, a sample would need to remain relatively still, and the tip would need to move relatively quickly. As Lev Manovich observes, one implication for images produced using rastering is that "It is only because the scanning is fast enough and because, sometimes, the referent remains static, that we see what looks like a static image" (100). However, atoms do not slow down enough to become referents; so, following Manovich's explanation, rastering would not work in the STM unless rastering is combined with the tip-surface interactions that provide the "stability" of a measurement of movement. Even so, STM researchers often need to correct for "drift" (or, when the sample atoms move before the tip has finished scanning) and also, at times, slow the sample atoms down by lowering the temperature to, for example, four degrees Kelvin so that the STM can scan the atoms.[31]

The raster scan allows STM users to convert tip-surface interactions into a camera of sorts. In so doing, the STM does not beam electrons (like a television camera) to assemble an on-screen image; instead, the STM forms what could be called a haptic camera, a "camera haptica" (playing on "camera obscura"), as the STM converts a series of interactions between atoms in the vacuum into a series of spatially arranged data points. The use of interactions as measurements moves the visualizing process closer to the sense of touch than the sense of vision or, more accurately, in a merging of the two senses—to haptic vision—because the tip gathers data about local interactions over a series of contiguous spots, and then presents the interaction data in a two-dimensional matrix, an image form. (The concepts of "camera haptica" and haptic vision are discussed further in Chapter 2.)

The rastering movement plays a role in the design of experiments, as researchers engage in practices that rastering affords. For example, Kandel and Weiss, in the experiment mentioned above, record images "with the tip rastering quickly along different directions, and . . . see a correspondence between this 'fast scan' direction and the locations at which atoms in the cluster either attach or detach" (8103). Kandel and Weiss present four versions of the same atoms that have been scanned in different ways, and use these versions to explore atomic properties (8104). Kandel and Weiss's experiment design is one example of how the tip's movement forms an experimental tool, allowing researchers to interact with surface atoms through the STM.

The ability of the tip to move up and down in the z direction also affords manipulation, because the tip can measure the three-dimensional electronic or topographic qualities of the surface. The ability to move in the z direction has made the distance between the tip and the sample a critical component of the operation of the STM from the beginning: Soon after Binnig and Rohrer developed the STM, before they achieved atomic resolution, Binnig and Rohrer "had to struggle with resolution, because Au [gold, their sample] transferred from the surface even if [they] only touched it gently with [the] tip" ("From Birth" 398). In 1987, R. S. Becker, J. A. Golovchenko, and B. S. Swartzentruber repeated this "mistake" of touching the surface with the tip; as a consequence, the atoms moved. In a letter published in *Nature*, Becker, Golovchenko, and Swartzentruber reported "atomic-scale modification" of a sample surface after they applied voltage to the tip. They attributed the modification to the transfer of a tip atom to the sample surface (421). Others repeated the experiment of using voltage pulses to "pin" molecules and atoms to a surface (J. Foster 29–32). In another experiment with the STM, for example, R. C. Jaklevic used the STM to dent a piece of gold, and then re-scan the same sample to measure how quickly the dent filled itself in (six to nine atoms per minute), thus using the STM as a tool to make the event of atomic movement measurable (Jaklevic 659).

The x, y, and z directional dynamics structure what can be seen with the STM. The STM images events through measuring change and movement, much like other visualization technologies; PET scanning, for example, images dynamic processes. The combination of x, y, and z directions that intensifies multi-directional interaction dynamics also affords researchers opportunities to manipulate individual atoms. For example, in another experiment, researchers J. G. Kushmerick et al. moved a nickel atom in order to observe how far (and how) atoms hop from place to place (2983). However, until Eigler and Schweizer published the article in *Nature* with the "IBM" images, researchers had not announced the manipulation of atoms by actually picking them up, "dragging" them, and repositioning them—a practice that the ability to move in x, y and z directions made possible. Although large-scale or automated manipulation of individual atoms is not a common use of the STM, even as I write this, manipulation of the surface becomes almost encouraged due to the dynamics created by the microscope's data-gathering process.

The coordinated x, y, and z directions also structure possible interactions between user and atoms through the arrangement of atomic interactions. The focus on measuring a series of interactions—and the ability to intervene, to change the interactions without destroying the sample—all encourage users to participate in the data-gathering process, as the measuring process becomes part of the experiment. The STM, then, is both a data-gathering probe and an experimental probe—and also an image processor, as explained in the next section.

GUI Image Processing Dynamics

The data produced through coordinated tunneling, rastering, and z-direction interactions is arranged in a matrix to create an image of atomic phenomena. Following the general process of scientific visualization, the raw data is arranged and thus transformed into images that viewers can interpret (see, for example, Brodbeck, Mazza, and Lalanne 30–31). The computer makes image creation relatively easy; however, despite the fact that Binnig and Rohrer worked for IBM when they developed the STM, they did not use computers to develop or operate the first STMs. Instead, Binnig and Rohrer created the first STM images by mentally building up atomic visions using an oscilloscope monitor's two-dimensional trails of the tip's rastering (Mody, "Intervening"). For a time, Binnig and Rohrer resisted the computer when they first began presenting their work. The first public STM image, outside of their own personal visions, was created from oscilloscope lines that Binnig and Rohrer traced onto cardboard, cut out, and glued together to form a three-dimensional model of the sampled surface (Binnig and Rohrer, "From Birth" 401).[32] The history of STM image development suggests that Binnig and Rohrer participated in imaging processes that encompassed or paralleled computer imaging capabilities, but did not entirely stem from use of the computer. (Chapter 3 further discusses imaging practices beyond computer-aided visualization, in relation to pictorial conventions.) However, Binnig and Rohrer switched over to the computer to generate images for publication. Other STM users followed suit when arranging their data, as the GUI (also developed in the 1970s, when the wave of non-lens-based scientific visualization technologies mentioned above occurred) afforded extensions of the interactive dynamics initiated by the data-gathering operations of the STM.

Characteristics of GUI also structure interaction, affecting what is imaged and how the resulting image is communicated. The fact that GUI use is now ubiquitous in scientific and medical visualization technologies (as well as in other technologies) makes the interaction dynamics of the GUI seem almost invisible; however, the interaction dynamics the GUI encourages help trace the influence of the GUI in STM use and in the production of STM images. The GUI affords STM users interaction with on-screen visual objects to manipulate in order to explore, change, or image the data. Interaction can affect the visual objects, as well as the data, with which the user engages. As information visualization researchers Dominque Brodbeck, Riccarde Mazza, and Denis Lalanne explain, with the use of computers (and the GUI), "graphical objects are not static anymore but can be interactively manipulated and can change dynamically" (29). In GUI interactions, the user expects to respond to visual objects (such as elements of images, or icons) as behavioral cues for manipulation, not solely in order to understand their meaning, or signification (Drucker, "Reading Interface" 215).

GUI characteristics affect STM use as well as image processing, as the STM user interacts with the GUI screen image in multiple ways, including while conducting the experiment, interpreting data from the experiment, processing the image for publication, and processing the image further if the image is intended to function outside of scientific journal articles (such as in press releases or on research group web sites).

The user's interactions with other dynamics, such as x, y, or z direction, allow the on-screen image to operate as an interface, thus allowing the image to function as an experiment and to help marshal evidence. GUI interactions structure the space in which the user interacts with the image, but also create a space in which the user interacts with the data and through the image with the nanoscale. The range of possible interactions that GUI enables reinforces the multi-directional and manipulable affordances created by electron tunneling, raster scanning, and z-direction moves.

Images and Experiments

GUI interactions also extend the time the researcher spends with the image and the data, further intensifying the imaging process. The arrangement of data in an image on a computer screen allows the user to turn the image into an experimental interface, coordinating with and amplifying the multi-directional affordance of the tip-surface interaction, or into what Daston and Galison call the "image-as-tool" (414). For example, in one of the experiments mentioned above, Jaklevic used images created by the STM to monitor an experiment. As interfaces, the STM images Jaklevic produced to observe the behavior of gold became events through which the researchers conducted the experiment. The involvement of Jaklevic and other STM users with the image (and the data) was intensified by incorporating the physical actions of the experimenters into repeated scanning and image constructions in order to conduct the experiment. The "Quantum Corral" image (Figure 4) is also a good example of how STM use becomes an event, as Eigler and other researchers created the nano-corral structure in order to conduct experiments on the electron standing waves they "caught" inside (Crommie, Lutz, and Eigler 218–220). The ability to use the STM image as an experimental space—to build structures (as in the case of the corral) or to observe events (as in Jaklevic's experiment)—is part of what makes the STM a significant visualization technology. As Gimzewski et al. comment in the journal Surface Science: "Scanning probe microscopies (SPMs) . . . , and in particular scanning tunneling microscopy (STM) . . . , have revolutionized the real-space imaging of molecules, providing a detailed understanding of the ways in which they interact with each other and with adsorbents" (101). Gimzewski et al.'s observation highlights the focus on using the STM as a scientific tool for understanding interactions.

The STM encourages further interaction with the data through the image, as researchers engage in understanding the significance of the data. Amann and Knorr Cetina explain the mode of practice that involves assessing significance as one in which scientific visuals "act as a basis for *sequences of practice* rather than observation at a glance. They [visuals] are subjected to extensive visual exegeses, rendering practices which attempt to achieve the work of seeing what the data consist of" (90). GUI characteristics suit STM images well to this interpretive task, thus extending users' interaction with the images. As one scientist I interviewed explained,

> It's [the experiment is] very much like putting something on a surface, seeing what it does, and trying to figure out exactly how it's behaving, and there's a lot of control experiments that you do between changing biases, changing tunneling currents, changing processing of the sample, to confirm like exactly what's happening, how you're looking at it.[33]

The ubiquity of the GUI, and the ability of researchers to use the same GUI to continue to interact with the image and explore and arrange data, while also making sense of the data and then producing images as evidence, all allow the user to go back and forth between Amann and Knorr Cetina's observational, interpretive, and evidence-producing modes of practice. While microscope users have almost always manipulated samples being viewed to produce images, the GUI intensifies and structures the involvement of the STM user in all stages of image production (Keller, "Biological" 110). The STM user's involvement creates a different relation to the image than if, for example, the user positioned the sample in an electron microscope and created an image, because the sample is destroyed in the process of viewing through an electron microscope, making only a limited amount of interaction is possible.

The STM user's involvement in image processing also encourages users to engage in continuing dynamics with the image and atoms; not only is it possible to engage with the sample again, but it is also possible to engage with the image again: The electronic screen affords the possibility of "refreshing" the image, just as the fact that the sample is not destroyed in STM scanning affords the possibility of "refreshing" the data through another raster scan. The GUI amplifies the effect of dynamic, manipulable images, data, and atoms. GUI dynamics thus structure interactions that invite multiple encounters, increasing a sense of immersion that heightens the feeling of engagement, as Rafaeli and Sudweeks explain. The affordance of the STM that allows users to repeatedly interact with samples through the image also reinforces engagement with the data—also perhaps suggesting the individuality of atoms through that repeated interaction.

Emphasis on manipulation as opposed to observation alone enters discourses about how the STM and its images can be used. For example, in an article describing imaging with the scanning tunneling microscope that appeared six years after Binnig and Rohrer won the Nobel Prize for developing the STM, IBM physicist John Foster writes, "After imaging a molecule, the next step is to do something to it" (26). Discourses about imaging functions also emphasize manipulation. More recently, a 2006 National Academy of Sciences Board on Chemical Sciences and Technology report on challenges and possibilities for chemical imaging acknowledges expanded uses of images that allow scientists to "do something to" what they see (Board 14).[34] Among important challenges for imaging, the report lists "understand[ing] and control[ling] complex chemical structures and processes"; "understanding and controlling self-assembly"; and "understanding and controlling complex biological processes" (Board 22–25). While the Board's report does not focus entirely on the STM, the report includes the STM as one of the visualization tools that can help researchers meet these challenges (114–21).

Producing STM Images: Image Processing and the STM User's Role

The interactive dynamics of the STM extend to the STM user's interpretation of the data, as well as the production of images as evidence through the tools of image processing. A key point of image processing, as John Russ explains in his *Image Processing Handbook*, is that "image processing, like food processing or word processing, does not reduce the amount of data present but simply rearranges it" (xiii). Russ's mention of arrangement in relation to image processing highlights one of the affordances of the GUI that becomes significant in structuring STM dynamics. While researchers using non-digital imaging processes may plot a graph from numbers or photograph experimental results, non-digital imaging processes limit how much researchers can change the graph or photo after production without also changing the data. In contrast, the process of digital image production associated with the STM allows researchers to be involved longer with the image during and after data collection. Prolonged involvement includes interaction with the data during what Amann and Knorr Cetina articulate as the transformation of data into images for publication that function as a "way of visually reproducing the sense of 'what was seen'" (114).

Like other digital imaging practices, the image production processes of the STM also incorporate the user into GUI practices that are contingent on interaction. Arranging information in visual form, and in the form of pixels, allows the STM user to continue the imaging process for far longer than a developer's involvement with optical film, extending the time the researcher participates in the imaging process. Michael Lynch's explanation of an extended imaging process in his study of the digital image productions of astronomers provides a sense of what also occurs with the STM images: "The real-time work of digital image processing involves a play at the keyboard, where images on the monitor are continuously recomposed by changing the palette, using touch-screen routines, plugging in parameters, and trying out different software manipulations" ("Laboratory Space" 72). The extended time that digital imaging processes require allows researchers to continue interacting with the data through the image and with different imaging techniques to develop images that contain experimental evidence.

To prepare an image for publication, researchers interact with the image to "clean it up," often by filtering the data. In an article that appeared soon after *Nature* published Eigler and Schweizer's images, in the *IBM Journal of Research and Development*, E. P. Stoll explains that raw data needs to be processed further due to interference, or "noise"[35] (following Shannon's division of information received into two categories, signal and noise), including noise that creates stripes "visible in nearly every real STM picture" (69). Therefore, some manipulation of the image, what Stoll calls "picture processing," tends to occur (87). Some STM researchers do not present filtered images in journal articles; however, "cleaning up" data is a common scientific visualization practice (Brodbeck, Mazza, and Lalanne 31).[36] Various ethnographers have discussed data processing as part of scientific image production. For example, Lynch explains data processing in astronomy images stating that raw data are "not treated as a pristine reflection of 'reality' but as the residue from a confused field where electronic noise, detector defects, ambient radiation, and cloudy skies mingle indiscriminately with the signal from a source object. The processed image is often considered the more accurate and 'natural' rendering" ("Lab" 70).[37] The impetus for "cleaning up" data, then, derives from larger habits of scientific visualization, and also contributes to further encouragement of interaction.

To filter data composing the image, STM researchers can use multiple techniques that further structure users' engagement with the data. For example, the data might reflect "drift" as the sample shifts over the time it takes for the raster to scan the surface: researchers may correct for drift, and then need to crop the picture.[38] Indeed, filtering is so important that scientists Sutter et al. have presented a way to filter the image at the level of data recording, through using a semiconductor STM tip that limits electrons in certain energy ranges even before getting to the image form (166101). In this case, scientists further incorporate the imaging software within the experimental apparatus, and blend data gathering and image processing. Alternatively, a researcher might use filters to enhance the contrast between light and dark, or to smooth out the contrast between different sections of the sample to see details. As one scientist explains,

> [I]n terms of daily usage, we generally . . . use other image manipulation techniques like taking the derivative of the surface. So then if you have a step [a point at which two uneven planes on the surface join like a stair step or terrace], if you take the derivative of the step, it just shows as a spike where the step is. So essentially, the contrast associated with having two different terraces at different heights goes away and so now those terraces appear like they're at the same height.[39]

Filtering techniques allow researchers to sort through data in order to begin interpreting the data: the researcher quoted above continues, stating, "these are actually the terraces, the same terraces we saw before, but here they're all a uniform gray color now. And now you see that these patches, which is actually what we're studying, you can clearly make out what actually turns out to be the atomic resolution in the patches."[40] Other filtering techniques include using Fourier transforms that show the frequency range of the data, helping to make the data measurable. Another researcher explains, "So if you look at a silver atom lattice [the structure the atoms create], I can get the periodicity of that [by taking the Fourier transform], take the inverse of that, and it gets me back to real space, and it will tell me that my lattice spacing is five Ångstroms."[41] These and other techniques allow researchers to highlight what information they consider important or to focus on more specific details such as the size of phenomena, for example. As researchers engage in highlighting data, they change how the image looks, and yet do not alter the data set. STM users may even process images further for cover slides for presentations, journal cover images, or for other scientific imaging contests beyond scientific papers.[42]

During the imaging process, the researcher also draws on other judgments and experiences so that the images created are, as Stoll comments, "aesthetically pleasing and informative and convincing" (76). Deciding to use color demonstrates some of the dynamics involved. False color has become a component of quite a few STM images, especially those appearing on journal covers and web sites (see Chapter 3 for more on color; also see Hennig, "Changes"). Many researchers apply color to highlight differences among pixel values. For example, researchers can highlight the three-dimensional appearance of the surface through assigning color- or gray-scale values to the various heights; researchers can also simulate illumination to introduce shadows or shading (Stoll 72). About the color choices in one image, one scientist explains,

> It [color] certainly can help to clarify the presentation because you can accentuate contrast in regions . . . which contain the point you're trying to make rather than have the reader be distracted by contrast related to things that you aren't worried about right now . . . . [In this image, for example,] you see all these lines here for atomic steps on the surface. So, if you look at that in black and white, your colors, your gray-scale has to be stretched to accommodate all those steps, and these steps are actually larger than any of the usual features on the surface. So some kind of stretching of the scale of the image has to be done in order to see the fine features that generally you're interested in. . . . So, if you look at the image just in gray-scale, it doesn't appear so clear because all the contrast is taken out by the steps rather than by the little bumps on the surface that you want to study.[43]

Adding color to STM images helps viewers grasp slight variations in value; Russ explains that human eyes can only pick out about twenty to forty shades of gray in an image, but can differentiate between hundreds of colors (35). Russ also observes that colors help humans verbally refer to parts of an image through different colors, as opposed to different shades of gray (35). Color allows viewers to not only distinguish differences in value, but also to import the value differences into language—another medium.

While added color helps viewers verbally and visually distinguish an image's particular characteristics, and so eases the entry of the presented value into scientific discourse, the researcher cannot rely on a color correspondence or order while choosing colors to use. Edward Tufte explains, "Despite our experiences with the spectrum in science books and rainbows, the mind's eye does not readily give a visual ordering to colors, except possibly for red to reflect higher levels than other colors" (154). Therefore, adding color to an image is dependent on the user's or group's decisions and previous associations with color. In his study of the digital image production of astronomers, Lynch explains that the dependence on individual or group associations to assign meanings to colors can lead to the user's reinforcement of her or his expectations of the sample through color choice: "Color becomes iconic when used for color enhancement or when signifying intensity or red shift. That is, the code is selected intuitively to suggest properties the object *should* have" ("Lab" 71).

In images of the nanoscale, the question of whether color correspondences exist is unsettled. Science studies scholars Arie Rip and Martin Ruivenkamp observe that color choices are not fully determined in the nano-researcher community: Some researchers see some colors as commonly used for certain features or attributes and cite, for example, the default colors in imaging software (29). Figuring out color schemes involves the researcher's preferences. One scientist I interviewed commented, "[Y]eah, so there's some playing that goes on with false coloring and just looking at what highlights the features that you want to show. Those [colors] aren't—of course, those aren't real."[44] Another commented that he would try out different colors and would make a file of those images, images he would later return to, and choose a version he liked best.[45] Using color, then, engages STM image-creators in another set of interactions to determine how to best present the data to users. Like other filtering techniques, choosing color extends the process of interacting through the GUI; choosing color engages the user in practices that also include scientific and extra-scientific cultural elements and conventions. By changing the appearance of the data and the image, chosen colors can also affect how data is read.[46] As color choices are not based on a predetermined order whose meaning viewers will immediately understand, but instead are based on previous (and most likely unexamined) color associations as well as the experience of color in the image, color choices affect the viewer's response to the image and his or her perceptions of what the image depicts.

Habituated Interactions: Coordinated Dynamic Effects

The operating dynamics of electron tunneling, movement in x, y, and z directions, and GUI use are separately identifiable in STM operations as well as in the dynamics of other visualization technologies. In the STM, however, the coordinated contributions of electron tunneling, movement in x, y, and z directions, and GUI use allow an intensification of the kinds of multi-directional interactivity and manipulability afforded individually by each dynamic. In the process of operating, these three dynamics constitute an instrument that encourages manipulability and interaction—interaction with and through images and data (as one might expect with GUI) , but also with atomic phenomena. As part of the coordinated dynamic of interaction, the user's involvement intensifies, allowing for an increase in the user's feelings of engagement (Rafaeli and Sudweeks). The affordances created by the coordinated dynamics of electron tunneling, movement in x, y, and z directions, and GUI in turn create rhetorical possibilities that are, in part, fueled by the feeling of engagement: manipulation and interaction themselves become rhetorical. Thus, manipulation and interaction form a strand of persuasive possibility that helps structure a user's experience within the STM, and also engagement with other technologies whose dynamics may operate similarly.

The increased engagement encouraged by the operating dynamics of the STM resonates with the last type of human computer interaction, "flow," further explaining how rhetorical experience with the STM is structured. As mentioned above, flow tends to include participation from both sides so that neither computer nor user occupies the positions of "sender" or "receiver." "Flow" may most accurately describe how the human user interacts with the screen and microscope apparatus because users engage in repeated interactions as they manipulate the sample and the image—interactions that build on responses from the sample, user, and image, and also take on a playful quality. Researchers' descriptions of using the STM echo the characterization of flow. Scientist Gimzewski and artist Victoria Vesna explain in a collaborative article, "through images constructed from feeling atoms with an STM, an unconscious connection to the atomic world quickly becomes automatic to researchers who spend long periods of time in front of their STMs. This inescapable reaction is much like driving a car—hand, foot, eye and machine coordination becomes automated" (11). In describing the experience of using an STM to someone who has never used the microscope, one scientist I interviewed responded by making an explicit link to computer games:

> Well, it's kind of like a late-seventies video game . . . . You're looking at a computer screen and you see little blobs on the screen which correspond to, you know, typically will correspond to a single atom or molecule on the surface that you're looking at, and you know, on good days you can manipulate these atoms or molecules, and then it becomes a lot more like a video game because you actually—most software interfaces are mouse based. If you want to manipulate something, that involves usually moving the mouse and clicking and then moving it somewhere else and clicking again. So that's almost like, you know, Ms. Pac Man or something like that . . . . So it can be quite fun.[47]

Another respondent answered, "it's hypnotic, I would say . . . . You go in and you're exploring a part of the world that nobody has seen because of that scale and you often don't know what you're gonna find and you usually don't understand what you're seeing."[48] Other scientists, such as Eigler, remark on the playful qualities of exploration with the STM and also on some of the excitement they feel as they manipulate atoms (Eigler, "From the Bottom Up" 425; also see Chapter 4). For these users, the hypnotic, game-like effects of using the STM may not only influence their perceptions of what they are doing (e.g., making STM use more "fun"), but also how the researchers perceive themselves and the instrument. As N. Katherine Hayles explains about her experience with virtual reality:

> I can attest to the disorienting, exhilarating effect of the feeling that subjectivity is dispersed throughout the cybernetic circuit. In these systems, the user learns, kinesthetically and proprioceptively, that the relevant boundaries for interaction are defined less by the skin than by the feedback loops connecting body and simulation in a technobio-integrated circuit. (*How We Became Posthuman* 27)

Whether or not users experience feedback loops, as Hayles describes, or a merging with the machine, Hayles's comment shows how users may attune themselves to operating beyond the confines of what is considered the boundary of the body and, in a transformative, prosthetic relation, merge or fuse—physically and mentally—amidst the dynamic that constitutes the interaction. The habits of interaction learned while using the STM may be longer lasting; as Hayles comments, while learning how to interact in a virtual reality simulation, "the neural configuration of the user's brain experiences changes, some of which can be long-lasting. The computer molds the human even as the human builds the computer" (*How We Became Posthuman* 27; see also Hayles, *How We Think,* Chapter 4). The interaction, the "flow," may then create longer-lasting patterns and, as with Hayles, above, Manovich and others have pointed out that the habituated patterns may also affect how the user responds to other interactive situations and technologies beyond the STM (136). As Manovich comments, for example: "As we work with software and use the operations embedded in it, these operations become part of how we understand ourselves, others, and the world" (136).

Indeed, one benefit of studying specific interaction dynamics that reach beyond the particular instrument, such as electron tunneling, movement in x, y, and z directions, and GUI use, is to see how the dynamics that structure STM interactions may influence trends in instrument use, such as the various uses of computer media—and through computer media, other scientific and medical visualization technologies. The fact that interaction dynamics such as rastering and GUI are also common to other technologies supports the claim that they may indeed become habituated interactions, so that one expects to manipulate, to interact with atoms, much like one expects to interact with an onscreen digital image. The dynamics thus import their own rhetorical power to the use of visualization technologies of which the dynamics are a part. It is no surprise, perhaps, given the presence of dynamics such as electron tunneling, movement in x, y, and z directions, and GUI use, that the STM includes such an emphasis on manipulability and interaction, but what is unique is how these operating dynamics combine to help create the particular space of flow in which users are persuaded of the manipulable, almost tangible, atom. The coordinated operating dynamics encourage specific interactions from image viewers in order to read them, as Chapter 2 and Chapter 3 discuss. The dynamics also affect the productions of the STM in ways that become important to understanding the rhetorical possibilities and changing discourses of nanotechnology in which atoms become tangible and able to be manipulated.

Interaction and Envisioning: Images, Information, and Atoms

In addition to the effects on the user discussed above, the combined dynamics of electron tunneling, movement in x, y, and z directions, and GUI use that help constitute the STM also help shape its productions, such as the image produced by the STM, the concept of information implied in this process, and nanoscale phenomena—such as atoms—that become visible through the imaged information. The following sketches trace the effects of the dynamics described above through productions of the STM in order to elaborate on what is persuasive about the STM, and to show the significance of the dynamic constitution of the STM. The following discussion also articulates aspects of the space of visibility created by STM use, and suggests how STM dynamics may affect our understanding of the nanoscale and of our world, as well as how we may have also changed in order to see atoms and molecules.

Imaging Interaction: The STM Image

In presenting data derived from electron interactions in an image form, the STM makes visible characteristics of the nanoscale; in the process, the STM also makes visible a concept of what an image can express and what an image can do. As other chapters discuss, the STM image contains particular rhetorical possibilities and challenges for STM researchers who use the images to communicate about the nanoscale, its possibilities, and challenges that may be shared by other digital technologies.

As discussed above, the STM image is composed of data. As with other, non-lens-based digital visualization technologies, the composition of the image from data significantly affects what the image can show or represent. Measurements of atomic interaction reflect electronic or topographical properties: STM scientists Joseph Stroscio and R. M. Feenstra conclude that viewers should consider STM images as descriptions of the "surface electron wave functions instead of atom core positions" (104–105). What may appear as a bump in an image may instead indicate dense electron activity, either as part of an atom or as a bond between atoms. Therefore, STM images do not often present an object:

the data does not necessarily compose a picture of a surface as the surface might look if we could snap a photograph. Instead, the image becomes a space in which collected data about movement and interactions coalesce. For that moment, what becomes static (and so becomes visible) is the placement of the data about a series of interactions in a matrix. Imaging with the STM makes visible data about surfaces and objects that are themselves composed of movement, from the sampled surface and tip's electrons, to the various chemical processes occurring in the sample at the nanoscale. Thus, it also becomes possible to visualize atoms and molecules *as if* those atoms and molecules were objects—to constitute a space of visibility in which atoms and molecules seem to stand still, to be not only grasped mentally as an object, but (as M. F. Crommie, C. P. Lutz, and Eigler, as well as Eigler and Schweizer show, for example) to also be grasped physically and moved.

The dynamics that visualize atomic movement and interaction become more intense due to the form of the digital image, whose production processes invite extended interaction with the data so that researchers can also move in among the data in the imaging process as well as afterwards, in image processing—or, in Amann and Knorr Cetina's terms, to use all three modes of practice that involve visuals, from manual and instrumental enhancement, to data collection, and to the creation of visual evidence (88). The affordances of the GUI encourage use of the same dynamics in experiment and image processing, allowing for images to function as interfaces with the nanoscale when the images are parts of the STM apparatus—and also for images to continue functioning as interfaces with the collected data, even when the images are separated from the computer that is part of the STM. As Daston and Galison observe, "Images [including STM images] become tools like other tools, part of the apparatus" (414). The shift to using images as interfaces or tools affects the appearance of the images and their use in contexts beyond the STM apparatus. Regardless of whether the images are viewed as part of the STM apparatus, or appear in data files or even in journals, STM images become an extension of the STM instrument in how the images allow—and even invite—viewers to interact with the data. The digital image becomes visible as an interface.

STM images invoke a more active viewer in that the combination of data and digital image invites viewers to engage with the image in order to discern atomic phenomena. The digital image, then, also becomes visible as a scientific instrument-interface, either on a screen as part of the

STM in the laboratory, or as it circulates in journals or data files. The instrument-interface as image suggests a shaping of knowledge about the nanoscale, one in which interaction with the image is emphasized in addition to the ability to manipulate and interact with nanoscale phenomena. The image form also suggests a possible challenge in understanding the nanoscale: The image is a form we most readily associate with showing objects, a tension that Chapter 3 further analyzes with regard to imaging conventions. The image form also suggests a more tangible relation with information as a way of understanding the nanoscale.

Information and Practices of Envisioning

The data that composes the STM image also forms attributes of the microscope; in the interactions that form the STM, the qualities of the data not only shape the image and the objects of scientific knowledge, but also reflect on the very concept of information. Contrary to the conventional and highly influential idea of information as a material-free pattern that corresponds to Shannon's theory in The Mathematical Theory of Communication, [49] the concept of information operative in the STM apparatus appears to be less abstract and more tangible as the data becomes visualized. A comparison between how information following Shannon's theory would appear in STM images and more embodied conceptions of information such as Donald McKay's, along with Susan Oyama's ontogenetic approach, highlights the differences between abstract and more tangible ideas of information. The differences provide a basis for understanding what STM images convey and how the images communicate that informs analyses of the significance and rhetorical effects of the images.

One consequence of using Shannon's information theory as a way of understanding the information produced by the STM manifests as a tendency to disregard the materiality of the image form, further implying that the interaction between viewer and image should also be unaffected by form or by physical changes or constraints; information should pass effortlessly between the image and the user (and vice versa), unless the information is interrupted by what Shannon calls "noise" (Shannon and Weaver 24). Shannon's theory also claims that information is not inherently connected with meaning in order to ascribe the same value to information that may occur in different contexts. Shannon's concept also leads to seeing the image as merely a pattern. The supposed ethereality of images on the computer screen supports the reading of images according to Shannon's theory, as we are accustomed to the flow of words or images as plays of light and shadow that lack permanent, onscreen traces—what hypertext theorist Michael Joyce implies is like writing with water (186–87).

STM images may seem to lend themselves to the understanding of information based on Shannon's theory because STM images, like other digital images, participate in a flexible citability: As Gilles Deleuze observes, digital images are in themselves citable, and not "original," as they are composed from a data set that can be expressed in many different ways, and so can create many different images, each as valid an expression of the data as the other (*Cinema 2* 265). Additionally, each moment brings a possibility of refreshing the image, of replacing the current view with another data set from another scan of the sample.

However, the fact that the process of making nanoscale phenomena visible includes so many material and human interactions suggests that Shannon's conception of ethereal communication may not best describe the way in which information about the nanoscale becomes mobilized and then communicated through STM dynamics. Instead, another understanding of information, one that builds on the theories of Shannon's contemporary, McKay, may offer a better descriptor of the "information" that emerges as part of STM productions.

McKay argues for a definition of information that includes the entire process of making meaning, from production of communication to reception of communication, and that, as Mark Hansen explains, "establishes (human) embodiment in its entirety as the *context* that determines what information will be selected in a given situation" (*New Philosophy* 79). Following McKay's view, the context in which the information occurs is important in order to determine an understanding of information. Oyama presents a related point in her account of the ontogeny of information that speaks to the role of the image as a form. Oyama argues that information does not merely exist by itself, outside of any form; instead, information co-develops ontogenetically with the form (2).

McKay and Oyama's views of information as not only material, but also as embodied in forms that co-develop with information, correspond to the interaction category of "flow" in the back-and-forth that creates the information. Understanding information as material and embodied also aligns with descriptions from researchers in using the STM mentioned above, including with the experience of interacting with the interface of the image in order to manipulate atoms and molecules. Atoms and molecules become more tangible as the nanoscale phenomena respond to the movements of the STM user. In this case, the STM user acts in context with information through the image form, and so builds up an experience of information as tactile, responsive, and understood through interaction.

What Oyama's insight also suggests for STM images is that the presentation of the data in an image not only changes the form, but the form itself also carries a certain number of determinants in the constraints and possibilities inherent in the form, therefore shaping the development of information. For example, the image form can communicate information in different ways than other forms, especially as the image is able to affect viewers—and affect viewers viscerally—before we understand the meaning of the image, as Steven Shaviro argues (26). As form and information co-develop and interact with each other in the process of formation, the information as well as how the information is presented are affected in this developmental process. Elements of "form" and "information," then, are inextricably part of the same process, and so what is gathered through visualization is substantially, materially different than what is gathered through the experience of information in a different pattern, such as a list of numbers or a graph.

When considering the workings of the STM, STM images, and the interface function in light of the different understandings of information presented by Shannon, McKay, and Oyama, Russ's simile in which he compares image processing to food processing (xiii) demonstrates one of the main problems with the prevailing theory of material-free information in regard to how we understand what STM images convey and how we learn about the nanoscale (and in so doing, demonstrates a limitation of the simile). While indeed the data may be able to be arranged, processing data in a particular way causes the information to collectively undergo a qualitative change, creating a different appearance that then generates different effects on viewers of the image. While one might be able to detect the same data in both cases, and while the data does remain the same, the viewer experiences different effects with each appearance.

The material qualities and the qualitative changes STM images undergo in image processing—like adding color or simulating lighting—create effects that become important when considering some of the functions of the images, including the information presented, although the data set itself may not change. Considering the special communicative qualities of images as a form that Shaviro explores, such as the ability of images to affect us—and affect us bodily—before we understand what the images mean, considering the materiality of images is especially important (26). Accounting for the material, embodied components of images is crucial in the case of STM images, because image production and even the STM-use process require the user's physical response to the images in the interaction, in the act of interfacing that is the image.

Assuming that information is material-free can lead to further assumptions that form is not important, or that differences in presentation (such as whether data about the nanoscale is presented in graphs or tables or digital images with false color) are also not important. Assumptions that information is material-free can lead to some ways of understanding the functions of images and rhetorical work in scientific discourses; however, assuming that information does have a material component—especially in relation to form—opens more avenues for analyzing and understanding what STM images convey and how the images communicate. The fact that different concepts of information, such as Shannon's, McKay's, and Oyama's, are in play also highlights a possible disparity in how images can be read and understood as part of scientific communication, as Chapter 3 explores.

Making Atoms Visible: Objects of Scientific Knowledge

The interactive operating dynamics of the STM outlined above help make atoms and other nanoscale phenomena visible through the STM: The understanding of information that also emerges provides some of the structure that makes atoms visible. However, returning to one of the questions with which this chapter began, how did the concept of atoms include the assumption of interaction for scientists? While the scope of this chapter does not allow for a full answer to the question of how atoms became manipulable, my analysis in this chapter provides a partial answer through attention to the visual expression of atoms. Hennig shows that from 1980 to 1990, changes occurred in the design of STM images that emphasized the concept of individual atoms appearing as building blocks. As Hennig explains (in the context of another image), the view of individual atoms in the image does not match the researchers' knowledge of the behavior of atoms, but instead seems to represent atoms ("Changes" 157). Hennig suggests that the reason for changes emphasizing individual atoms in STM images is that the visual design of images was aligned with a common vision of nanotechnology, as being able to build from the atoms up. Through their visual design, the images thus helped make the STM an important tool for nanotechnology (Hennig, "Changes" 143). Hennig's argument is sound and important—and the shift in understanding atoms as expressed in STM images, for example, may have been swayed by such visions of nanotechnology. However, the evidence presented in this chapter, from the analysis of STM operating dynamics, including the effects on the STM user's experience with the instrument, suggests that other influences also may affect how the images are composed and what the images express about atoms. For example, the deeply interactive processes that let experimenters experience the movement of atoms, and that create affordances for manipulation, may shift the visual expression of atoms. Additionally, the presentation of data in images affects the overall appearance of the images, as what also becomes visible is a different understanding of data as highly manipulable aggregates, as fluid—yet also as solid, as bounded objects because of the computer's ability to create images that present the data using, for example, representational techniques such as shadowing, false lighting, and perspective. (Other chapters in this book discuss the techniques listed here.)

Conclusion: Fields of Visibility, Possibilities for Communication

> *Technologies, like rhetorics, themselves, serve as interfaces for human relations and endeavors.*
>
> —Stuart Selber

Regardless of whether or not a connection exists between our ability to image atoms with the STM and the more concrete understanding of atoms that emerged in the 1980s, atoms become visible as entities that can be individually imaged. However, the parallels between scientists' shift towards thinking of atoms as manipulable and STM development are interesting and suggestive. The interactive practices whose coordinated dynamics help constitute the STM, from electron tunneling to raster scanning to the graphic user interface, in addition to STM productions—image, information, and concept of the atom—also make visible nanoscale phenomena in ways that render atoms and molecules as tangible entities. As constitutive elements of the STM, dynamics such as electron tunneling, movement in x, y, and z directions, and GUI use, create interactions that emphasize movement as much as material; solicit longer (and perhaps more involved) interactions with atoms and with data in image form; and measure local interactions as well as arrange the measurements in a matrix. Use of the image form to communicate also carries something else along with the data presented in a matrix of sampled data points: a reification of the atom and an accompanying, insistent rhetoric of atoms as manipulable, individual entities. However, as this chapter's exploration of what is made visible through the STM also points out, STM users also occupy different positions of visibility within the new configuration that is the STM and the images produced by the STM. In other words, the STM constitutes the space of visibility for atoms that seems different from what those early quantum physicists expected, in part because we have changed in our interactions with information in the process of using the STM as well as in our interactions with the images produced by the STM. Such transformations have a few effects that are worth exploring further.

In the process of transforming the haptic interactions among atoms and between atoms and researchers into the realm of the visible, seeing becomes a way to access senses other than sight—the computer screen converts data into light that is then assembled on the computer screen as an image. Thus, the STM makes visible not only movement and interaction, but also touch—through the measurement of interactions to the manipulation of atoms and data. (Chapter 2 expands further on the visibility of touch.) As the rhetorics of the STM include the process of manipulation and interaction (as well as the promise of that interaction), the STM also persuades through an embodied, haptic process that itself participates in these rhetorics, one that is not likely limited to the STM, given the presence of operating dynamics common to other visualization technologies, such as rastering and GUI use.

Adapting Stuart Selber's statement quoted above, this chapter's analysis of interaction in STM operating dynamics has also yielded an identification of how the instrument as well as the image functions as an interface for human-atom relations and endeavors. Analysis of the STM and image as interfaces is significant for understanding the rhetorical processes and practices that are more persuasive in STM use. For example, in exploring how the space between electron clouds becomes an interface, a site of interaction, we see the ways in which manipulating atoms became not only possible, but almost impossible not to do given the z-direction functioning of the STM. The invitations to interact with atoms that the z direction affords users, found even in the process of creating the tip—and to respond to the actions of the atoms—encourage a persuasive encounter. The persuasive interaction that the z direction encourages is constitutive, and occurs while the STM visualizes atoms as part of an experiment—or, in other words, as researchers are engaged in the process of knowledge-making, of determining the significance of the data. Chapter 3 and Chapter 4 further explore the rhetorics of interaction while analyzing conventions and tropes; however, the focus on interaction helps explain how the idea of manipulating atoms forms a topos both at the STM level and in arguments about the significance of nanotechnology. As a topos, "interaction" also becomes significant for how it functions as a touchstone for an experiential *practice*, a way of knowing more about atoms, as opposed to a touchstone for a concept or a thing. Further, analysis of interaction as a topos suggests the importance of studying the rhetorics of experiences or procedures in the STM, its images, and other scientific and medical visualizations.

The fact that the dynamics that compose the STM are shared by other visualization technologies suggests that another value of studying the STM is as a way of marking where we are in the development of visualization technologies. The focus on common operating dynamics as an analytic object suggests further assessment of the procedural and of the participatory rhetorics at play in interactive technologies such as the STM. The quality of flow that characterizes the form of interaction suggests further exploration of affective and haptic rhetorics of visualization technologies.

Use of STM images to convey information about the nanoscale contains challenges to communication: Stroscio and Feenstra (among others) point out that STM images do not show objects, but instead show electron intensities that may or may not correspond to surface topography, thus presenting images that look as if the images actually do contain objects. Images such as the "Quantum Corral" image (Figure 4) can therefore create assumptions of actual objects, and may lead to misunderstandings of what nanotechnology can be. Chapter 3 and Chapter 4 discuss such challenges in more detail in regard to reading images. First, however, Chapter 2 discusses another category of cultural, historically specific practices that STM users engage in when they use the microscope, and that those viewing STM images also engage in: vision practices.

2 Camera Haptica: Blindness, Histories, and Productions of Haptic Vision

Vision and its effects are always inseparable from the possibilities of an observing subject who is both the historical product and the site of certain practices, techniques, institutions, and procedures of subjectification.

— Jonathan Crary

While explaining how the scanning tunneling microscope operates, D. M. Eigler employs a simile that suggests how much the STM apparatus differs from instruments that create optical images, and also suggests how much STM users depart from vision practices used in optical image creation:

> Incongruously, it [the STM] forms an image in a way which is similar to the way a blind person can form a mental image of an object by feeling the object . . . . Just as the blind person can sense not only the shape of the object, but also its texture, compressibility and thermal conductivity, the STM can tell us much more about a surface than just its shape. ("From the Bottom" 427)

In linking two practices that do not use optics to create images—image formation by a microscope and mental image creation by a sight-impaired individual—Eigler uses anthropomorphism to equate a microscope's process of image formation to that of a human's, bringing the atomic images produced by the STM closer to human scale. The simile creates a parallel to the imaging process of the STM: as the tip of the microscope travels over the surface and interacts with atoms, the STM builds the data matrix from which the image is produced, using an action similar to a finger brushing along a line of Braille or a weave of cloth. In addition, the capability of the tip to move up and down and with z-direction movement allows the STM to sense other attributes of the surface, expanding imaging possibilities of the STM to include touch. In a similar use of the simile, scientist James Gimzewski and artist Victoria Vesna state the importance of expanding imaging possibilities enabled by the STM. For them, the STM "represents a paradigm shift from seeing, in the sense of viewing, to tactile sensing—recording shape by feeling, much like a blind man reading Braille" (11).

Eigler re-uses and expands the simile a few sentences later: "Even more remarkable, just as the blind person can push, pull, pick up and put down objects on a table top, so too can we push, pull, pick up and put down surface atoms using the tip of a microscope" ("From the Bottom" 427). By adding "we" to the already-linked microscope and seeing-impaired person, Eigler entwines the physical apparatus and STM users by linking the imaging process of the STM with the viewing practices of microscope users, thereby erasing the usual separation between machine and human. The metaphoric fusion of machine and human imaging processes in Eigler's simile indicates more than a mere turn of speech. As scholars from a range of disciplines argue, metaphors affect not only how we envision ourselves mentally and physically in the world, but also how we envision the less physical, but no less real, bodies of knowledge we create.[50] The references to blindness raised by Eigler and Gimzewski and Vesna reflect how the STM user may conceptualize her or his vision practices in the interaction necessary to use the STM, and the use of the trope suggests that the process of envisioning nanoscale phenomena with the STM is profoundly imbricated with the machine apparatus. The trope also suggests that the process of using the STM involves a visual practice that departs from our usual understanding of vision—so much so that the observer is comparatively blind.

Eigler's figurative comparison of what a STM user can learn from how the microscope produces images to the sensory capabilities of a seeing-impaired person is also a literal comparison if we employ a conventional understanding of "vision" as optical, as incorporating perspective that relies on the perception of a tableau or image at a distance seen by the eyes of a viewer. Instead, the vision practices that Eigler references use non-optical methods to image and manipulate atoms, practices that fuse touch and vision.[51] The vision used with the STM, haptic vision, merges two senses usually defined as separate; the viewer does not engage in visual practices like those of perspective that rely on a separation of the senses.[52] In fact, the imaging processes of the STM and its user discourage the use of perspective because the distance between the microscope tip and sampled surface does not function as a clear or immaterial conduit, as air does in perspectival vision. Instead, as Chapter 1 discussed, the space functions as a contact interface due to electron tunneling, and so constitutes part of the microscope's interaction.

The merging of Eigler's STM user and instrument, of STM user or image viewer and surface atoms, can be described as a blurring that blinds STM users and image viewers to their own perspective while observing with the STM. This is a blurring in which STM users and viewers forget that they are observers. Paul de Man refers to a similar blindness when he suggests that being blind to the workings of language permits the critics he discusses to achieve their insights: "their language could grope toward a certain degree of insight only because their method remained oblivious to the perception of this insight" (106). In the case of the STM, the method of perception creates a blindness in which the atom and the atomic scale become visible—and from which emerges different possibilities for the observing subject that Jonathan Crary refers to in the epigraph to this chapter. In turn, the possibilities for observers help create a different dynamic than one where optical vision dominates. In the dynamic of seeing with the STM, blurring or blindness becomes an integral component.

Moreover, because the practices through which Eigler figuratively entwines the STM, its users, and the visually impaired differ from practices leading to more conventional optical vision, the use of haptic vision creates effects that have implications for understanding how and what the STM and its images communicate. The practice of haptic vision affects the rhetorics operating in the use of, and in the images produced by, the STM by creating different relationships between observer and observed and, consequently, subject and object. The interplay between observer and observed, and between subject and object in STM vision practices thus articulate different relations between image, user, and viewer or audience.

How haptic vision and optical vision affect relationships between viewers and objects differently appears clearly in comparison to Aristotle's articulation of the three ways a speaker needs to reach his or her audience, commonly conceptualized as a "rhetorical triangle." Briefly sketched, the writer or speaker at one corner of the triangle sends a message (another corner) to the audience (the third corner) about a topic, considering how to present him- or herself effectively and appeal to the particular audience, modifying the presentation of the message accordingly. Articulations between message producer, message, and audience are much less clearly identified in the relationship created by the STM apparatus and STM images. The user of the microscope is creator or "writer" as well as viewer or audience member; the user's STM use relies on responding to an image that then may alter the image and/or the sample under view.

Similar to how hypertext collapses writer and reader, the position of "writer" and "audience" fuse at times, and the "message" does not remain separate because the message changes in response to the writer or audience due to the participation of the writer/audience in the imaging process. Hence, the positions of viewer/user and audience blur in relation to each other and to the "message." Blurring between user and audience causes the rhetorical strategies developed in accordance to the fixed positions of the rhetorical triangle to not necessarily hold sway. Like the possible effects on rhetorics in hypertext,[53] the blurring between STM user and audience produces possibilities for different means of persuasion as well as different moments where such rhetorics function more intensely, such as within the imaging interaction itself.

Haptic vision affects scientific communication particularly because one major stake in science is exactly what constitutes an object of study for a discipline. How an object of study is viewed helps shape its constitution. William Ivins argues in his frequently-cited *Prints and Visual Communication* that through perspective, the "exactly repeatable pictorial statement" that was a main tool of the scientific revolution gave scientists the ability to create consistent images of objects by following perspectival methods of depiction (23-24). Visual consistency allows scientists (and others) to understand how a representation of an object corresponds to an actual one, and also to recognize objects viewers have seen in images, as well as images of objects viewers have seen (Ivins 23-24). Therefore, if scientists engage in visual practices that do not rely on perspectival conventions (such as haptic vision), and produce images that also do not use perspective, then concerns about the reproducibility and authenticity of scientific knowledge become pertinent. Understanding not only how images are produced, but also how images create and are created by visual practices become significant for understanding science and scientific communication.

While scholars in film studies, psychology, and others writing on visual culture describe other instantiations of what can be called haptic vision, both in present and past cultural contexts,[54] exploring haptic vision in scientific practice can more fully describe the rhetorics at play in verbal and visual scientific discourses. While rhetoricians of science have examined visual, as opposed to only verbal or textual rhetorics, studies of scientific rhetoric tend to focus on the representational elements of images.[55] (This is discussed further in Chapter 3.) So far, rhetorical explorations of the fusion of touch and vision have been restricted to studies of gesture.[56] Studies of gesture do not fully articulate the interactions such as those produced by the production processes for STM images. As the comparison with the Aristotelian rhetorical triangle reveals, considering how STM images persuade given the transformative and transforming vision processes that are necessary to create and view STM images can inform other studies in visual rhetoric, the rhetoric of science, and science studies.[57]

While Chapter 1 focused on how atoms are made visible through the deeply interactive practices researchers engage in to use the STM apparatus and image the nanoscale, this chapter explores another part of how and what use of the microscope makes visible—the practices of haptic vision out of which STM images emerge and the transformations that practices of haptic vision enable. How do users of the STM, as well as viewers and users of the images the STM produces, engage in vision practices that deviate from optical practices so much that observers become "blind?" I argue that the interactions that constitute the operating dynamics of the STM, such as electron tunneling, movement in the x, y, and z directions, and GUI use, also encourage the fusion of the STM user's tactile and visual senses. Haptic vision differs quite dramatically from the optical vision necessary for perspectivalism and the consistency that the "exactly repeatable pictorial statement" conveys: haptic vision thus re-articulates the positions of observer and observed. Examining more specifically what haptic vision is, how haptic vision has emerged, and how haptic vision operates within the context of STM use can further explain the dynamics of haptic vision and the dynamics of images of the nanoscale. Because haptic vision not only describes a use for eyes that departs from perspectivalism but also forms the basis for investigating new communicative rhetorics and practices such as interaction (as discussed in Chapter 1), this chapter shows how haptic vision affects the formation of scientific knowledge and rhetorical practices.

Camera Haptica: Haptic Vision

But what happens when what you see, even though from a distance, seems to touch you with a grasping contact, when the matter of seeing is a sort of touch, when seeing is a contact at a distance? What happens when what is seen imposes itself on your gaze, as though the gaze had been seized, touched, put in contact with appearance?

—Maurice Blanchot

The contact at a distance that Maurice Blanchot describes evokes how STM images affect observers through a fusion of sight and touch: an image leaps out from a computer screen, journal article, or cover, and grasps at the eyes of the image observer, possibly causing her or his fingertips to tingle with the feeling of the surface. Whether the observer then responds by running her or his eyes along the surface, or by manipulating the STM apparatus to move atoms, she or he undergoes a synesthetic experience, touching the shimmering surface while knowing that the image does not show a representation, that the colors are not real, and that the atoms do not really exist as solid, tangible entities due to their characteristic movement—and yet, the atoms seem to be *there*. Steven Shaviro reads Blanchot as also suggesting "that the image is not a representational substitute for the object so much as it is—like a cadaver—the material trace or residue of the object's failure to vanish completely" (17). In their scans of the fleeting positions of atoms, STM images tug the observer back to her or his world of physically experienced knowledge as the images blur the boundaries between touch and vision, vibrating with the material traces of atoms, residues of interactions among electrons, the bright colors grasping the eyes: The observer experiences the image bodily, the observer's gaze suddenly transformed, "seized, touched, put in contact with appearance" (Blanchot 75).

Blanchot's questions also apply to the experience of using the STM, judging from Eigler's comparison of the experience of using the STM to that of a seeing-impaired person using a cane, as well as other descriptions of the process of the STM. For example, in a *Scientific American* article, Gerd Binnig and Heinrich Rohrer describe the viewing process as allowing the observer "to 'see' surfaces atom by atom" ("The Scanning Tunneling Microscope" 50). Binnig and Rohrer do not attempt to explain the operations of the microscope to the non-specialist, and often non-scientist, readers of the magazine in terms of a fixed object—an overall surface or individual atom, for example. Instead, Binnig and Rohrer suggest a process of moving from one atom, one location, to another, more like touching each atom one by one in its place. In addition, Binnig and Rohrer use a different simile to convey the visual contact the microscope employs: "trying to probe atomic structures with visible light is like trying to find hairline cracks on a tennis court by bouncing tennis balls off its surface" (50). The intermingling of touch and vision also occurs in Binnig and Rohrer's Nobel acceptance speech; they recount their "first attempts at chemical imaging," describing gold "islands," or clumps of gold atoms, in terms of words that evoke vision as contact: "the islands were visible as smooth, flat hills on a rough surface in the topography" ("From Birth" 399). The chemical imaging Binnig and Rohrer refer to does not show the topography of the surface, but instead creates images of the electronic properties of the surface; yet, Binnig and Rohrer describe their first STM image in terms of topography—as rough and smooth, qualities most apparent to fingertips, not eyes. The emphasis on texture also appears in professional journal descriptions of imaged atoms as "protrusions in the image" (Kandel and Weiss 8103) or "protrusions centered in depressions" (Pascual et al. 12632). Such descriptions of STM images from scientists articulate a viewing process that includes bodily movement and physical connection with atoms, a process that incorporates touch into vision—into both the images the STM produces and the experiences that STM users and viewers of STM images undergo. The fusion of touch and vision is accomplished through the eyes and often the fingers, and so becomes a process of haptic vision.

Perhaps in part because the different spatial qualities of "when seeing is contact at a distance" often take the forefront in articulations of the haptic, most scholars in art history or film, as well as other theorists who discuss the haptic, point to Alois Riegl's articulation in his *Late Roman Art Industry* as a starting point. Riegl describes humanity's cultural development in terms of understanding space in two different ways: the optical and the haptical. Haptic space, in Riegl's terms, is a space of surfaces that often includes decorative surface formations such as patterns of rugs or woven fabrics. The observer explores haptic space by moving from one discrete object to another, where each object is separate and separately experienced—as opposed to optical space, in which the observer can survey objects at a glance and determine the spatial connections of objects. Riegl's constitution of the haptic as space has been taken up by film theorists like Antonia Lant and Laura U. Marks, who jettisoned Riegl's teleological cultural development vector and used his consideration of the spatial characteristics of the haptic as a jumping-off point for their arguments. Riegl's haptic space is also useful to describe STM images, as the spatial qualities of STM images are one of their main distinguishing characteristics, as Chapter 1 discussed.

While Riegl's space is also useful for discussing the process of haptic space, Gilles Deleuze and Félix Guattari's description of the haptic in *A Thousand Plateaus* emphasizes the viewing process, and so speaks to the imaging dynamics of the STM. Starting from Riegl, Deleuze and Guattari articulate haptic space not only as the product of a culture-wide vision practice, but also as a mode that occurs in the experience of smooth space, a space emerging out of certain relations between an observer and the environment. Smooth space allows an observer different movements and operations than the striated space where optical vision operates:

> The first aspect of the haptic, smooth space of close vision is that its orientations, landmarks, and linkages are in continuous variation; it operates step by step. Examples are the desert, steppe, ice, and sea, local spaces of pure connection . . . one never sees from a distance in a space of this kind, nor does one see it from a distance; one is never 'in front of,' any more than one is 'in' (one is 'on' . . .). (493)

Deleuze and Guattari articulate the dynamics of haptic vision on a continuum, whose opposite pole is optical vision where, at a given time, a viewer may be engaged in varying degrees of one or the other, separate or intermixed. Moving away from Riegl and his followers, space becomes less of a constant, and Deleuze and Guattari suggest that no sense of background is possible. The absence of the concept of background alters how an observer not only sees, but also responds to the environment:

> Where there is close vision, space is not visual, or rather the eye itself has a haptic, nonoptical function: no line separates earth from sky, which are of the same substance; there is neither horizon nor background nor perspective nor limit nor outline or form nor center; there is no intermediary distance, or all distance is intermediary. (494)

Instead of backgrounds or limits as orientations, then, Deleuze and Guattari suggest that the "step by step" movement of passing between points is the primary characteristic of the smooth space of close vision. Haptic space does not include background or a sense of self that exists apart from the haptic space. This consideration of space illustrates some of the differences between optical and haptic modes of vision as well as differences between different theories of what haptic vision constitutes. The differences that emerge from optic and haptic vision include, then, not only what and how the observer sees, as Riegl suggests, but also, as Deleuze and Guattari explain, how the observer relates to the objects and the environment as well as actually—and actively—participating in that environment.

A comparison of the main model of optical vision, the camera obscura, with a model of haptic vision, what I call a "camera haptica," clarifies the different relations between perspectival and haptical vision. The camera obscura gained popularity in the late 1500s, and became entrenched as a model of vision in the seventeenth and eighteenth centuries. The camera obscura has been used for various purposes, including scientific and popular observations of the world, the study of vision, and artistic practices, and has been an acting "model, in both rationalist and empiricist thought, of how observation leads to truthful inferences about the world" (Crary 29). The camera obscura's apparatus is composed of variations of the following: an observer sits or stands within a darkened chamber, one wall of which has been punctured with a small hole from which light from the outside world beams onto the opposite wall and projects the scene outside. Despite many changes in the actual apparatus during the seventeenth and eighteenth centuries, what remained constant for the camera obscura is the relational constitution of observer and observed. The relations between observer and observed, Crary explains, configure subjectivity in a new way, one where an isolated, autonomous subject peeps out at the outside world (38). The observer position within the camera obscura also becomes "a figure for both the observer who is nominally a free sovereign individual and a privatized subject confined in a quasi-domestic space, cut off from a public exterior world" (39). The camera obscura model separates the act of seeing from the observer's body, and separates the observer from what is observed. A related example of this separation, described by Brian Rotman, is the emergence of the vanishing point within Renaissance perspectival painting. The vanishing point, the point at which the perspectival lines converge, is the point at which the observer stands, and in so doing, also stands in the place of the artist, therefore seeing from the artist's view.

Yet, the vanishing point is also "in a meta-linguistic relation to [the objects represented in the image], since its function is to organize them into a coherent image" (19); therefore, the observer is doubly removed from the outside world.

The relation of interiorized, bodiless observer peering at a representation of the outside world that the camera obscura establishes contrasts from the relation occurring in what we might call a camera haptica. In a camera haptica, the observer is fully embodied in the practice of haptic vision, as observations are not limited to vision alone, but often depend on the observer's bodily interaction. The observer is thus immersed in what she or he observes; the world and the observer fuse, and in so doing, transform both world and observer. The camera haptica does not include a darkened room; instead, the observer's nerve fibers, eyes, and skin, as well as the object and milieu being observed, become the camera haptica. The camera haptica *is* the ecology or environment within which the observation can be said to take place. In a camera haptica, the walls, and the space between the walls and the observer, comprise the object with which the observer interacts, becoming giant touch-screens or malleable surfaces allowing the observer to respond physically to the surface as she or he travels along that surface.

The space that the camera haptica articulates is not that which exists between observer and observed, but the space that the observer traverses *in the process* of seeing, the space the observer moves through as she or he runs her or his eyes across the atoms one by one, building a data map from one point to another, like the creation of a data map from sampled points on a surface using the STM. The observer may view an object, but the experience of the object, the movement from one object to another, takes priority as the experience of haptic vision draws the viewer in, blinds her or him (in a way) as the objects become visible, as able to be experienced. The communication dynamic here can be expressed in terms of Jacques Derrida's formulation in "Signature, Event, Context" that relies on a definition of communication as operating in the term "communicating doors:" What is conveyed is not signification necessarily, but is instead a shock or force, information not in terms of representation, but instead of contact (309). Haptic vision emphasizes event over object, action over stasis, and as such, there is no distancing objective observer possible, or a distanced object. With haptic vision, new and different possibilities emerge for conceiving of space and time as well as relating to objects and subjects.

Emergent Conditions, or Forces for Haptic Vision

> *Most of the historically important functions of the human eye are being supplanted by practices in which visual images no longer have any reference to the position of an observer in a "real," optically perceived world.*
>
> —Jonathan Crary

Just as the composition from different interactions of the STM can be traced to previous or concurrent dynamics used in visualization technologies, such as electron tunneling, movement in x, y, and z directions, and GUI use, the haptic vision practices associated with the STM also share attributes with several other vision practices. Although these visual practices may share similar characteristics, a direct, causal link or incorporation into a developmental narrative is not what interests me, because, as Crary argues against the common origin story that the camera obscura led to photography's invention, "Such a schema implies that at each step in this evolution the same essential presuppositions about an observer's relation to the world are in place" (26). Therefore, focusing on such continuities alone may obscure crucial differences in the visual practices of a particular time, as well as understandings and functions of observers.

However, looking at the vision practices related to a particular visualization technology in the contexts of individuals involved, the incorporation of instrument techniques, the kinds of image production practices, and contemporary cultural trends forms another way to understand a technology by suggesting affinities with other visualization technologies or practices. In addition, following the habituated patterns of relations between observers, other objects, and technologies reveals patterns of vision practices and articulates positions that observers can occupy. A few visualization technologies that have contributed to or paralleled the vision practices of the STM exhibit characteristics that repeat those vision practices, and so help to show some of the forces at work, as well as the conditions of the emergence of the STM. Before elaborating on these connections, however, my argument depends on a definition of vision as culturally and historically determined and determinable. I briefly rehearse a few arguments for vision and perception as having cultural components to explain how what is defined as "vision" can range through a wide array of possibilities over time and across cultures.

Cultural Components of Vision

The common view of perception, of light rays striking the retina and thereby creating an image of an object or the world outside, what perceptual psychologist James Gibson calls the "orthodox theory of the retinal image" (58), seems relatively straightforward, especially as this concept of perception describes the mechanics of light rays and the operations of the eye. However, on closer examination, the common theory of perception is less simple. For example, as Gibson argues, the idea of the retinal image does not fully describe how vision operates. Gibson reminds us that "natural vision depends on the eyes in the head on a body supported by the ground, the brain being only the central organ of a complete visual system" (1). Also, as Martin Jay observes after describing how eye mechanics function, while we understand how vision works mechanically, "the precise manner of its translation into meaningful images in the mind remains somewhat clouded" (7). Another occluding factor is that many related practices are often grouped with the "orthodox theory" of perception, including "the terms and practices of Euclidean geometry, theoretical optics, cartographic and graphic representation, the crafting of instruments and the artistic techniques of linear perspective" (Lynch, "Laboratory Space" 56). The effect of grouping related practices, Lynch observes, is that this "'orthodox theory' has given us an observation language that is ready to hand and all to easily relied on in epistemology" (56). Lynch refers to the grouped practices as "opticism," similar to what Jay calls "Cartesian perspectivalism" and what art historian James Elkins, in *The Poetics of Perspective*, comments has become the main version of perspective in recent times (xi, 217-261). More than a description of how the eye actually operates, opticism sets up a relation between a fixed observer and a separate object or image within a visual field.

Psychologists and other scholars who study vision have commented that at least parts of this orthodox theory, or opticism, are culturally based. These scholars argue instead that vision relies on movement and bodily experience. For example, Gibson argues against common conceptions of vision as what a stationary observer uses to receive information about the outer world. He explains: "[V]ision is kinesthetic in that it registers movements of the body just as much as does the muscle-joint-skin system and the inner-ear system" (183). Bodily experience as well as movement, then, is a component of vision: Gibson argues that vision operates by intermingling both proprioreceptive (inner) and "exteroreceptive" (outer) stimuli, not just one or the other. The act of vision involves much more of a bodily experience than that suggested by opticism. In addition, the separation of vision from the other senses or bodily states is also not necessarily fixed. As Brian Massumi observes, the co-mingling of senses occurs more frequently than we might assume. In fact, Massumi argues, "Vision always cofunctions with other senses, from which it receives a continuous feed and itself feeds into: hearing, touch, proprioception, to name only the most prominent" (145). The interdependence of the senses suggests a basis for synesthetic metaphors in describing sensory experiences that do not fit conventional definitions of sight or touch. Studies of vision such as Gibson's also imply that other kinds of vision are possible besides "opticism," and that other relations between observers and what is observed are possible.[58] The line between what an observer sees due to mechanics and what she or he sees—and senses—due to learned practices blurs at times; such smudged lines between vision and other senses suggest the possibility that visual practices proliferate, are not necessarily unchanging entities, and are not solely biologically determined.

Theories of perception and what constitutes vision have also changed according to cultures and periods. For example, philosopher Ivan Illich charts how the role of extramission (the idea that vision starts with a light ray that emanates from the viewer's eye to the object seen, and then returns to the eye) in ancient optics led to a different understanding of perception ("Scopic" 5-8). Art historians have researched a number of culturally and historically specific aspects of visual practices. For example, Crary explores the shift in the nineteenth-century observer through relating the nineteenth century's "'separation of the senses' and industrial remapping of the body," which Crary argues resulted in the division of touch and vision (19).[59]

Historical and cultural factors also affect the configuration of vision within individual perception, and so can also lead to different kinds of visible knowledge. A few examples in visual studies include Joel Snyder's "Picturing Vision," in which Snyder argues for a historically based view of realism that focuses on a particular understanding of the "object of depiction" as "a standardized, or characterized, or defined notion of vision itself" (223). Samuel Y. Edgerton describes the early, inaccurate attempts of seventeenth-century Chinese artists to copy European technical drawings, revealing the extent to which linear perspective was culturally based (273–287). Following Walter Benjamin, John Berger's immensely influential *Ways of Seeing* helped publicize the idea that vision is not a monolithic, unchanging entity, but is instead culturally influenced. Some critics, like Robert D. Romanyshyn and Marx Wartofsky, even suggest that all perception is cultural, produced by such objects as images within a particular culture.

In sum, one way of understanding what counts as vision is that vision is composed of not only biological mechanisms, but also of complex historical, cultural, and technological configurations. Since, as Foucault reminds us, every history is a history of the present, historical descriptions of possible vision configurations are interesting both for their historical specificity and for their articulation at this time, amidst such an efflorescence in different visualization technologies that, as many writers discussing the historical construction of vision have noted in passing, suggests that our current practices may be changing.

Instances from psychology and the history of vision practices suggest that historical and cultural conditions affect vision practices, and so create different objects of knowledge as well as different positions for observers. Three related practices occasioned by visualization technologies that also share cultural and historical conditions with the vision practices of the STM, and provide a context for the emergence of its vision practices, are microscope vision, informational vision, and the closely related digital vision. These characteristics also describe some of the conditions of the emergence of haptic vision found in STM use. Each of the three practices also include a certain amount of Blanchot's contact at a distance, where the observer is "blinded" as she or he becomes immersed in vision practices different from those constituting perspectival vision. Microscope vision, informational vision, and digital vision as practices deserve more involved treatment than I have space for here; therefore, what follows is a sketch that highlights the shared characteristics and conditions of emergence of the STM and haptic vision.

Microscope Vision: "A sincere Hand and a faithful Eye"

From the time of the development of the microscope in the seventeenth century, microscope users have engaged in and developed a set of vision practices with which to use the instrument, practices that form part of the context for STM-associated vision practices. One consistent characteristic of microscopes, as Evelyn Fox Keller argues in "the Biological Gaze," is that microscopes merge "looking and touching into an undifferentiable and unified act" (120). The merged act of seeing and touching present in microscopes occurs in part because, even with optical microscopes, perspective does not help viewers understand what they see through the microscopes' eyepieces. Since the early days of the instrument, microscope users have experienced optical illusions and distortions due to impurities in the glass lenses fashioned by lens-makers. Also, microscope viewers cannot verify that what is seen through the eyepiece is the intended object instead of a bit of dust, a reflection of viewers' eyes, or a part of the object under view. Given these inherent difficulties, Ian Hacking summarizes, "all but the most expert [observer] would require a ready mounted slide to see anything" (192). Instead, viewers wishing to see with microscopes must learn a series of practices that not only include seeing, but also intervening (Hacking 189). Hacking cites George Berkeley's *New Theory of Vision* (1710) as a source of this intervention, "according to which we have the three-dimensional vision only after learning what it is like to move around in the world and intervene in it" (189). Therefore, to see with microscopes, microscopists must develop practices that use both hands and eyes.

While microscopes have always been associated with both touch and vision, until recently, the role of touch has been played down in scientific accounts, because often what was emphasized was either the instrument as an "extension of the senses" (where "senses" stands in for the word "vision"), or the visual productions of the microscope. With microscopes, a reliance on not only vision, but also touch, is mentioned from time to time, but is not a main discussion topic in popular, philosophical, or historical work about microscopes. One reason for the lack of attention to touch is that the association of sight with knowledge, modeled in the workings of the camera obscura, also became a dominant association in science. The association of sight with knowledge also includes the denigration of touch. As Barbara Maria Stafford recounts, in the eighteenth century, for example, the untouched, unchanged specimen was prized; any sign of manipulation or manufacture was considered the work of charlatans, not scientists. Stafford states:

> "Objectivity," or the honest conduct of the practitioner, was thus synonymous with the absence of any visible sign of manufacture. The rise of objectivity as a scientific ideal in the early modern period was facilitated by the development of measuring and distancing apparatuses. These truly 'automatic' devices seemed to preclude shady handling and phony gadgetry. (103)

Scientists' focus on eliminating the human hand and possible interpretation of nature continued in other contexts. For example, Lorraine Daston and Peter Galison find reliance on mechanically-produced images in their study of late-nineteenth and early-twentieth century concepts of objectivity in scientific atlases. The convention of eliminating any signs of human manipulation in scientific work led to an emphasis on the visible, not manipulable, aspects of microscopy. The convention also parallels the campaigns of early modern science writers against ornamental language, and the adoption of a more "plain" prose style that, as its proponents argued, show the writer's modesty and dedication to "clear provision of virtual witness" through the writer's prose (Shapin and Shaffer 66). The connection of visual clarity, truth, and objectivity continues as a general assumption with implications for the status of the use of touch, as Keller summarizes: "In scientific discourse, looking is associated with innocence, with the desire to understand, while touching implies intervention, manipulation and control" ("Biological" 107).[60]

Nonetheless, vision and touch were intertwined from the early days of the microscope, and how both senses do so strikes parallels with the haptic vision sketched above. For example, in 1665, Robert Hooke noted that what a microscopist needed was "a sincere Hand and a faithful Eye" on the fourth page of the the preface to *Micrographia,* and described the proper method to reach "the true philosophy" (see for example the seventh page). The observer begins with her or his coordinated hand and eye to view the sample, moves to memory, to reason, and then back to the hand and eye, as Hooke details on the seventh page of the preface. The work of the hand includes not only the manipulation of the sample under view, but also the manipulation of the pencil or pen in drawing what the microscopist observes. Additionally, Hooke's method forms a loop, for the microscopist must view and draw repeatedly to create the final representation. Hooke reports on the twenty-fourth page of the preface:

> I never began to make any draughts before many examinations in several lights, and in several positions of the light, I had discover'd the true form. For it is exceeding difficult in some objects, to distinguish between a prominency and a depression, between a shadow and a black stain, or a reflection and a whiteness in color.[61]

The process of repeated viewing and drawing that Hooke recommends creates a visual practice that Michael Aaron Dennis calls "disciplined seeing" and which, he argues, standardized the viewing practices of microscope users (323). The benefit of standard practices was the development of a common visual and verbal discourse among microscope users (323). The process of "disciplined seeing" continued after the seventeenth century, as Keller recounts, describing intertwined touch and sight as "the great contribution the rise of an experimental ethos brought to nineteenth century biology: the desire—and increasingly the skill—to reach in and touch the object under the microscope, and to thereby 'make it real'" ("Biological" 112).

Keller's descriptions of the haptic and visual interactions with biological specimens are one example of the convergence of hand and eye in an interactive and haptic vision. Like optical or electron microscopes, the STM merges touch and vision in the viewer's examination of the microscope sample. However, the dynamics of the STM fuse the senses of touch and vision further due to the fact that STM users are encouraged to not only touch, but also to build at the atomic level—the fusion of touch and vision is integral to the operation of the STM, such as the tunneling functions and rastering of the tip, discussed in Chapter 1. The STM thus includes an even more intensive haptic feedback loop with the user in order to generate the data that constitutes the image. Also, the characteristics of the image allow the user to interact further with the sample through the image, both through the digital form of the image and through its visual conventions and tropes. (Chapters 3 and Chapter 4 explore this further.)

Microscope vision relies on not only the existence of the loop Hooke describes, but also the anticipation of the next step in the procedure of interaction, as it becomes a part of the vision practice. As Keller notes, "for the experimental biologist, the function of the microscope is not simply to enhance looking, or even to validate the tangibility of that which we are gazing at, but to employ that gaze as a probe in anticipation of action" ("Biological" 114). The probing gaze of microscope vision practices, and the anticipation the gaze includes, parallels the haptic qualities of the movement of the STM's tip across the sample, as well as the responses of the observer.

Informatic Vision

Another set of vision practices that provides context for the conditions of emergence of the STM and STM images are those used for reading information in visual form. Readers and composers of informational images engage in different reading and composition practices than those either producing text or representational images, as those working in information visualization observe. Edward Tufte observes, "To envision information . . . is to work at the intersection of image, word, number, art. The instruments are those of writing and typography, of managing large data sets and statistical analysis, of line and layout and color" (10). The practices of creating and reading data images encourage habits of interacting with data and images that differ from habits that may arise from interacting with images or text alone, thereby creating hybrid practices of both production and viewing.

Reading an informational image is more haptic than optical; the observer's eyes range amidst the data of the image, reading "atom by atom" or data point by data point. Haptic reading creates bodily effects because the reader engages in a dynamic not only with the information, but also with, following Susan Oyama's argument in *Ontogeny of Information*, the formation the data takes in image form. Reading images for data includes the tactile pleasure of moving over the data, and also the pleasant anticipation of where the future data touch—and, if on the screen, in the possible transformation of data points to provide further understanding of the data, thus continuing both pleasure and anticipation. The observer enters into a state of fascination, compelled to look and keep looking, running her or his eyes over the matrix of figures presented in vivid colors that are not tied to meaning-making, or signification, in and of themselves. The observer experiences communication before meaning through the affective experience of merging with the data points one by one, blind to the overall image as the observer haptically engages with the data.

The affect generated in response to informational images has been recognized indirectly at times. For example, since October 1999, the pop-tech magazine *Wired* has aptly titled its "Raw Data" department "Infoporn," and features wildly imaginative and colorful data graphics that seem designed to seduce the magazine's readers through color, shape, and gratuitous revelations of information. *Chartporn*, a blog created by economist and artist Dustin Smith, also indicates the implied affect in informational images. *Wired* and Smith's nods to affect support Oyama's argument about the co-development of information with form, as well as its inherently material registers: If the information were bodiless, according to Claude Shannon's prevailing cultural conception, information transfer in image form would not involve the somatic effects that the *Wired*'s section title implies.

Informational vision, then, includes not only hapticity, but also affect and bodily response, contradicting the cultural conception of information as bodiless and material-free. The function of STM images and content suggest the use of informational vision, although haptic vision also includes other elements, such as the physical response that allows for change in the image, facilitated by the computer.

Digital Vision

The vision practices associated with digital media, digital vision, include some of the important characteristics that compose the visual practices of the STM. The characteristics of digital vision—including the concept of data arranged in a visual pattern and the ability of that pattern to change in response to an observer's physical responses, thus changing the observer into a user—necessitate a separate discussion, as digital informational images encourage different responses than non-digital informational images. What is different about digital vision, most dramatically in its computer-enabled forms, is the viewer's ability to physically respond to the observed images and to have the expectation of altering the images. The haptic vision produced by the STM shares with digital vision a structural component of interactivity.

Lynch discusses the space in which digital vision occurs in his articulation of two modes of laboratory space. Lynch describes digital space as one "built up from a series of repeating digits or cellular units that act simultaneously as codes and substantive details" ("Lab" 63–64). In Lynch's view, the digital mode deviates from opticism in ways similar to haptic vision, including pixilation, in which each pixel can express a range of values following an arbitrary code; the ability to manipulate each detail; the composition of each image as not "a static field reflecting its object but a spatial product of a temporal scan;" and the finiteness of each detail, "limited by the array of pixels 'beneath' which there is no finer order of detail" (64). The characteristics of digital space allow for different and changing relations between image and creator as well as image and observer.

The computer enables users to develop practices in response to interaction with the computer screen. When computer programmers moved from programming with command lines to using graphical user interfaces (GUIs), interaction with elements onscreen and with other people through the screen began to seem possible, as GUI inventor Douglas Engelbart first suggested in a 1968 conference presentation. Engelbart explains how he developed the idea of direct, physical user-screen interaction and also the on-screen combination of text and images through a vision he had while working as an electrical engineer for what would become the National Aeronautic and Space Agency (NASA): "graphic vision surges forth of me sitting at a large CRT console, working in ways that are rapidly evolving in front of my eyes (beginning from memories of the radar-screen consoles I used to service)" (189). Engelbart's vision contains imagery that evolved "within a few days to a general information environment where the basic concept was a document that would include mixed text and graphic portrayals on the CRT" (189). Interestingly, Engelbart's memories of using one of the first computer-generated informational images, radar, led to the development of the GUI.[62]

An emerging area of haptic technologies, sometimes called sense-of-touch technologies, has recently intensified the haptic uses of the screen as a visual map. Developers of haptic technologies create tools for interacting with the computer through touch, often in conjunction with sight. For example, haptic tools such as the Phantom (a relative of the joystick) allow a user to feel properties of an object, enabling users to feel the contours and rigidity or elasticity of an object. Haptic technologies can be used to recreate experiences: In June 2003, using a dataglove and a Phantom, researchers at the University of Buffalo New York Virtual Reality Lab announced that they had transmitted over the Internet the sense of a particular object's touch that an experimental subject felt (OhAnluain). In order to feel, subjects had to follow the movements of others. The director of the Virtual Reality Lab, Thenkurussi Kesavadas, comments: "The big breakthrough for us was to figure out that the users needed to actively track and try to repeat what the other person is trying to do, and that was when the receiver started reliving what the sender was trying to do" (qtd. in OhAnluain). Through re-experiencing, the receiver (or second user) can follow the first user's experience. This invention from the Virtual Reality Lab parallels how the STM tip transmits the experience of the sample surface to viewers of the STM image through haptic vision practices. The invention also parallels the STM user's visual experience by including experiential response with perception.

Another area of research in computer-enabled digital vision, the development of virtual reality interfaces, links more directly with the STM. In 1993, Warren Robinett at the University of North Carolina, Chapel Hill and R. Stanley Williams at UCLA connected a virtual reality system to a STM to create a "Nanomanipulator," taking Eigler and E. K. Schweizer's atom manipulation a step further to make the interaction seem even more real for users. The system provides tactile feedback when the microscope is in the tip-moving mode (when an image cannot be produced) so that users can "feel" instead of haptically see across the surface. While the Nanomanipulator team's website notes that the interface is mainly used with the atomic force microscope, it still can be used with the STM as well as with other scanning probe microscopes ("The Nanomanipulator"). The team describes the Nanomanipulator: "A scientist using the Nanomanipulator can view incoming data, feel the surface, and modify the surface (using voltage pulses) in real time" (Taylor et al. 127). The point of the device is to "[approximate] presence at the atomic scale, placing the scientist *on* the surface, *in* control, *while* the experiment is happening" (Taylor et al. 127). The scientist can immerse him or herself bodily in the interaction even more than with a non-virtual-reality assisted experiment because the Nanomanipulator includes a force feedback handgrip, a head-mounted display that "provides highly detailed 3D color images in real time," and a head-tracker to change the graphics in coordination with the user's head movements (Taylor et al. 128).

As these examples of haptic technologies show, the vision associated with digital images that allows expansion into haptics relies on bodily interactions with the image. (Furthermore, Engelbart's inventions, the GUI screen and mouse, for example, have become familiar objects and actions for many first-world individuals: These examples of haptic technologies also highlight that in one common activity, computing, we do not always use optical vision practices). The user/participant in computer-assisted digital vision learns to respond physically; as a consequence, the user/participant develops habits of response.

The STM shares much with synesthetic vision practices such as microscope vision, informational vision, and digital vision, and as their existences suggest, these practices occur side-by-side with the dominant, scopic regime of perspective. What is important to study about vision practices is their intensity, frequency, and influences on the rhetorics at work surrounding their use and production, and how vision practices alter their users and perhaps jostle habits created by the dominant visual practice of opticism. The alterations enabled by microscope vision practices, as well as those caused by computer-mediated techniques and technologies, are far-reaching and pervasive across all media (Manovich 19). The three vision practices sketched briefly above also involve viewers' more intense participation, seen not only in STM use (as Chapter 1 discussed), but also in the use of other recent visualization technologies more familiar to rhetoricians, such as *Photoshop*, or *iMovie*, for example.

STM-Haptic Vision: Dynamics and Productions

One of the main aspects of perspectival viewing is encapsulated in how, in the fifteenth century, visual artist Leon-Battista Alberti famously envisioned the artist's canvas as a window-like plane that the viewer "looks through" to the picture. Alberti created a grid that an artist could use to divide what he or she sees into squares to draw if the artist immobilizes her or his head and covers one eye. In this way, the artist creates the perspective of what is seen through the entire grid. The resulting drawings thus feature a single point of view, with the viewer of the image standing in for the artist as he or she gazes at the scene. In fact, while Albrecht Dürer's famous engraving of an artist using Alberti's "window" includes only what looks like a pole to orient the artist, in another of Dürer's engravings of an artist at work, the artist's head is immobilized to keep the artist's view steady. Alberti's method of imaging a scene from a fixed perspective is not how the STM apparatus visualizes atoms, as Chapter 1 explains, nor is Alberti's method the one employed by practitioners of haptic vision with the STM. Like other visual practices—such as microscope vision and informational vision—the STM user engages in a practice of haptic vision in which the observer's eyes are no longer perceivers of distant, detached objects, mere senders of visual information to the conscious brain. Instead, eyes become points of contact for physical response as well as vision, reacting more like nerve or skin cells to a stimulus. Since, as Shaviro points out, images cause visceral reactions or responses before there is time to create signification, or meaning, the viewer's physical response does not necessarily travel through the conscious, knowing subject (26). The delay that Shaviro explains is also physical: As neurophysiologist Benjamin Libet shows, a delay occurs between a stimulus (such as a tap on the arm) and when the person becomes aware of the stimulus. Also, as Libet explores, the initiation of voluntary actions occurs at times before consciousness (35–45; 259–306). Events before consciousness form just a few of our unconscious responses, and includes use of habituated motor skills, such as those employed in driving a car while having a conversation or thinking of what to eat for dinner, for example. Also, as Libet points out, "intuitive thought patterns even when complex, creative thinking and problem solving are involved (as attested to by great mathematicians, artists, etc.)" (xxv). Events before consciousness also occur in digital image reading: In contacting the differentiations of pixels on the screen, for example, eyes and the computer screen interface with other parts of the body and the on-screen image; eyes that are attached to nerves and hands connect to microscope parts other than on the screen, forming a feedback or circuit, like the loop that Hooke describes is part of traditional microscope practices. Each component of the feedback loop communicates with the others, producing a vision practice where the end goal is not necessarily an image, but is instead part of the process of image production.

Operating within a dynamic fusion of what in perspectival vision would be called the observer and the observed, the dynamics of STM vision could best be described as mimetic, as the observer is engaged in a tactile contact with the image and also, in the case of atom manipulation (for example) with the changing image onscreen. Some elements transformed through the practice of haptic vision include dynamics within the observer; that is, the functions of the eye and body and the boundaries between the observer and the observed. As will become apparent below, while these different elements alter so much in their fusing that they are not really separable, categories familiar to English speakers accustomed to the dominant, Western scopic regime of Cartesian perspectivalism highlight the differences between perspectival vision and haptic vision of the STM. The specific mimetic relation STM users engage in affects the boundaries drawn between subject and object, and allows for new possibilities for the observer.

Dynamics Within the Observer: Responses to Haptic Vision

In "The Work of Art in the Age of Mechanical Reproduction," Walter Benjamin argues that while watching a film, unlike while viewing a fixed image like a painting, because an observer does not have time to contemplate or reflect on the meaning or associations she or he has with an image before being shown something else, images impress on the observer and create a visceral contact through which the film communicates, physically affecting him or her (238). This visceral shock, as film theorists following Benjamin such as Shaviro and Marks ascribe to the tactile qualities of film, is also apparent in haptic vision practices such as in STM use or in running one's eyes, atom by atom, over a STM image.

Mimesis creates what anthropologist Michael Taussig calls "a palpable, sensuous, connection between the very body of the perceiver and the perceived" (21) and, in turn, creates a dynamic that transforms both observer and observed. Taussig follows Benjamin's conception of the tactile work of images, and describes the transformation of both the "copy" (in the case of the STM, image, tip, person following the movement, etc.) and the "original" (atom, surface, atomic movement, etc.) through the tactile, habitual operations of mimesis: "The wonder of mimesis lies in the copy drawing on the character and power of the original, to the point whereby the representation may even assume that character and power" (xiii). Therefore, what may be transmitted in an image may not only be a copy of the atomic movements, but something of the atom's characteristics, or, in Blanchot's words again, the "material trace or residue of the object's failure to vanish completely" (83).

Like the operation of memes, patterns of information, thought, or behavior that are possible for one individual to "catch" from another like a virus, the contagious dynamic of the STM operates as part of a larger system, and not as an always-conscious decision on the part of the observer. Meme replication is not always caused by the meme carrier or because of recipient's conscious volition or agency; it occurs within a larger communicative network, as anyone who finds himself or herself passing along a rumor or using a slang interjection can attest to. The moment of contagion or mimesis operates as a moment of blindness; the observer does not act, but is instead acted through, becoming a conduit for the force of communication. As Susan Blackmore explains in regard to agency and memes, "we have to think of them as autonomous selfish memes, working only to get themselves copied. We humans, because of our powers of imitation, have become just the physical hosts needed for the memes to get around" (7–8). Both physical actions and the ways of seeing STM images on the screen—or later, in journals—operate mimetically: they are not created or envisioned by one person, but are communicated through her or him as well as through the image and pattern of interactions.

Like an image created from a matrix of interactions through the STM, haptic viewing is not one static moment, but an event composed of many moments, because the viewer's eyes must contact many points in order to create a series of responses that move, as Binnig and Rohrer phrase it, "atom by atom." Not only must the observer's eyes move from spot to spot to see fully, but her or his head and body must also move. This contrasts with optical vision, where the observer must hold still to see the image. In the case of STM manipulation, the observer's eyes often trigger a physical response that may include bodily movement to adjust the microscope tip in order to interact with the atom. The observer then engages in a mimetic process that highlights the viewer's bodily movement over time—again, quite the opposite of perspectival vision. Jay notes that the development of perspectival vision "led to a visual practice in which the living bodies of both the painter and viewer were bracketed, at least tendentially, in favor of an eternalized eye above temporal duration" (55). Such an eternalized eye in the use of haptic vision would be blind, as haptic vision depends on temporal and bodily experience.

Therefore, one of the main contrasts between the practices of haptic vision and perspectival vision is that the viewer reacts to images in qualitatively different way: Instead of viewing from a distance and at a standstill, as visitors to a museum gallery tend to gaze at a painting, the observer of STM-produced images sees by running eyes like fingers from point to point. The haptic reactions of viewers transform the habitual relations between different parts of the body in the process of observing, and so create possibilities for new capacities of experiencing, feeling, and acting.

Observer and Observed

STM images communicate with the observer in a mimetic process more like contagion than conscious adoption in both the presentation of data in STM images, over which the viewer runs her or his eyes, and in the changeability of STM data as the STM apparatus interacts with the surface, possibly altering or rescanning the surface in response to the viewer's responses. In addition, the STM includes other mimetic dynamics beyond the image, for the tip following the sample mimics the sample surface, causing the viewer to not only engage in haptic vision with the image, but to also haptically respond to a series of changing images as she or he follows anticipated moves of the atom in the process of manipulation.

The viewer's bodily movement and contact also transform the relation of categories that are conventionally separated, such as observer and observed. Because the haptic image through which atoms are glimpsed (the observed) does not immediately summon representation, the image forestalls any identification process. Instead, because of prolonged visual contact with the surface of a sample, the haptic image encourages a relationship between the observer and what she or he sees. As Marks comments, "it is not proper to speak of the object of a haptic look as to speak of a dynamic subjectivity between the viewer and the image" (*Skin* 164). While Marks suggests that the dynamic subjectivity that hapticity summons helps to create cultural, embodied memories in film, in the case of the STM or imaging programs like *Photoshop*, the haptic dynamic also includes a certain range of physical responses. The observer reacts within the haptic dynamic by altering the image—both observer and image change in response to each other. The dynamic is more like that of picture-making (i.e., painting or drawing). Elkins compares picture-making to the experience of blindness: "When an artist is concentrating, trying to feel the exact pressure of the lead and even the texture of the paper as the pencil skips across its surface, then vision is occluded" (*Object* 226). Physical interactions between the different elements take priority, causing a lack of attention to sight, and transforming what is "viewed" into what is experienced as the observer, like Elkins's artist, helps to create the image. While the priority of such interaction is not as explicit when an observer pores over images in a journal, for example, the characteristics of informational vision practices discussed above suggest that the viewer is still transformed as she or he must move her or his eyes across the image and engages in haptic vision to follow the data, contacting the image and data—and so, entering into a different relationship with the haptic image than with a perspectival image.

In the transformation of the relationship between viewer and viewed, haptic vision dissolves the boundary between inside and outside. This transformation is analogous to the qualitative shift that Elias Canetti describes, from an individual subjectivity to crowd subjectivity. Canetti explains that the shift to crowd subjectivity occurs in part because the individual who becomes part of a crowd loses the fear of being touched, the fear of transgressing a boundary (the skin), of the entering in of the outside. In the crowd, an individual's fear gives way to relief as the boundaries between inside and outside shift from that of the individual to that of the crowd (Canetti 16). Similarly, with haptic vision, the boundary between inside perception and outside view dissolves, in addition to the boundary between conscious and unconscious response. The individual melts into Canetti's crowd with what is viewed: The observer and observed become engaged in a relation qualitatively different from one where they are not in such contact. The relation caused by haptic vision is based on movement of the eye and the movement inherent in the mimetic give-and-take between the observer and the observed.

One example of how the mimetic give-and-take operates in atom manipulation performed with the STM comes from the same article as Eigler's blindness figure discussed at the start of this chapter. Eigler narrates his and Schweizer's development of their manipulative abilities at the nanoscale. In September of 1989, Eigler and Schweizer noticed streaks across STM images they had made of xenon atoms on platinum (Eigler, "From the Bottom Up" 430–31). Once Eigler and Schweizer realized that the movement of the xenon atoms had caused the streaks during the imaging process due to interaction with the tip, the scientists realized that they needed more capability to interact. Therefore, Eigler and Schweizer modified the STM so that it could switch back and forth between an imaging mode and a mode where they could freely move the tip across the surface (432). Eigler and Schweizer then flipped between one mode and the other, imaging one xenon atom and moving the atom. In their switching, Eigler and Schweizer combined action and vision in a way suggestive of Eigler's metaphor of a blind person—the scientists moved the atom by feeling its position, achieved within the combined experience of moving the atom and viewing the image the STM produced from the "blindness," or feel of the atomic interactions, of the STM itself. In the moment of blurring, Eigler and Schweizer extended seeing into a haptic situation in such a way that their positions as observers enlarged; they became users able to move the atom. In the blur of interaction, the scientists became extended, incorporated into the STM, allowing for the production of atomic manipulation.

That Eigler and Schweizer's ability to interact with atoms relies on hapticity, on shifting the observer to user and then fusing with the STM and atom, is further supported by the actions taken by the developers of the Nanomanipulator to heighten hapticity. Instead of attempting to visually image those moments when the STM is in movement mode, developers of the Nanomanipulator augmented the existing hapticity by introducing a "force feedback probe to allow the user to directly control tip motion during modification and to feel the changes as they are occurring" ("Nanomanipulator"). Seeing and moving an atom occurs through hapticity, not through optical or perspectival means.

Eigler and Schweizer's image-viewing process was also based on what the process could tell them about their previous and future physical moves, as the scientists viewed an atom in the imaging mode less like an object than a map. Therefore, Eigler and Schweizer's response to the streak and subsequent actions was not based on observing an image for its representational qualities. Instead, the scientists felt for the way the atom moved, following or imitating the imaging mode with the moving mode and creating habits of interaction that led to Eigler and Schweizer's understanding of what their interaction processes allowed. Through interaction, and because of the possible transformation of both observer and observed, the observer and what she or he sees become immersed in responses. In addition, the haptic viewer becomes a relay, not a receiver, as she or he may become in contemplating a representation. The position of the observer as a relay is also situated between the current moment and the anticipation of the future—of future moves and possibilities—as the observer also brings to the experience the anticipation that he or she will generate a response that may end up changing the object of knowledge as it changes the observer. As he or she changes, the observer becomes a place within a new kind of visibility in order to see objects like atoms with the STM. Therefore, as the boundary between observer and observed smudges through vision that is also in contact, the body of the observer becomes involved and affected; as the action occurs, the relation between image or screen and observer also becomes one of habitual interaction, interaction that transforms the individual by creating one entity from individual and image.

Productions of STM-Vision

Because of the merged relation between observer and observed, what has been ascribed to the position of subject and object of vision in such systems as Cartesian perspectivalism blur and, through contact, transform. In addition, an observer's adoption of the transformative subject position available in haptic vision produces a different subjectivity than that commonly associated with the humanist subject and its subset, the scientist subject.

Subject and Object

Haptic vision highlights the fusion in which subject and object are merged. The merged relation of subject and object affects what becomes visible as scientific objects, as STM-produced images show: STM images describe scientific objects seen through image series (such as Figure 1) or scientific objects obviously built through interaction with atoms (Figures 1, 2, 4, 5, and 6, for example). In STM images, "objects" become events in their action and reaction, like the "IBM" series or R. C. Jaklevic's image series mentioned in Chapter 1. The objects and images form part of a larger interaction that incorporates the STM user, instrument, and practices of haptic vision. The object is not known only from one angle, or known from qualities such as shape, color, or size in relation to the environment of the object, cues that optical perception provides. Additionally, the object that becomes visible with the STM—a pattern of electron interactions, a concentration of electrical activity, or a sudden gap in surface topography—is also known through hapticity, as topography becomes important in the interactions of the microscope's tip, and as the resistance of an atom to being moved across a nickel surface, for example, becomes apparent to the STM user through the instrument adjustments that the user makes in response.

One effect of haptic vision practices is that haptic engagement can expand understanding of knowing and interacting with an other. For example, as she reveals a stake of understanding the subject and object differently in terms of film, Marks explains that

> [h]aptic cinema, by appearing to us as an object with which we interact rather than an illusion into which we enter, calls upon this sort of embodied and mimetic intelligence. In the dynamic movement between optical and haptic ways of seeing, it is possible to compare different ways of knowing and interacting with an other. (*Skin* 190)

How one knows about atoms, then, becomes embodied, experienced, and more than observed. In addition to Marks's observation about such comparisons, the exploration of haptic vision practices alone also shows how we might develop capacities for envisioning an other in ways that are literally not objectifiable; for, as subject and object fuse and are transformed, the usual line between them blurs.[63]

Productions of Subjectivity: Possibilities for the Observing Subject

> *In a haptic relationship our self rushes up to the surface to interact with another surface. When this happens there is a concomitant loss of depth—we become amoeba-like, lacking a center, changing as the surface to which we cling changes. We cannot help but be changed in the process of interacting.*
>
> —Laura U. Marks

Haptic vision affects STM users, data, and atoms as they contact and respond to each other, and in so doing, transform. Therefore, the act of haptic vision changes those possibilities for the observing subject, to use Crary's term in the epigraph for this chapter. Because of changes in possibilities for subjects, for example, determining a separation between elements such as observer and observed becomes impossible, just as Gregory Bateson reminds us that it is impossible to speak of a separation between the seeing-impaired person and cane while the person is using the cane (64). In relation to visual models, a user of the camera haptica posited earlier in this chapter is one who is completely implicated in the camera haptica, and who partly determines the form of the instrument as she or he becomes part of the instrument.

The regular interactions the STM user and viewer of STM images engage in during the operating dynamics such as electron tunneling and GUI use, described in the Chapter 1, become habitual. Habits such as GUI use or haptic vision practices are neither only mental nor only physical. Elkins provides one example of how habits of seeing manifest: After describing how an artist draws in a way that seems to be non-optical, he turns to an effect of such a process:

> It is easy to tell an artist from an art historian by the way she responds to a drawing. A historian, trained with books and color slides, will stand at a respectful distance and look without moving. An artist, at home with gestures, will want to move a hand over the drawing, repeating the gentleness of the marks that made it, reliving the drag of the brush or the push of the pencil. The drawing has *become* its bodily response, and the body moves in blind obedience to what it senses on the page. (*Object* 227)

Elkins's art historian employs a vision that does not need to involve the body's response other than the circuit from eye to brain, where the endpoint is the brain, and a vision that needs distance between drawing and observer, like with perspectival vision. In contrast, the artist's response more closely follows that of the visually impaired person in Eigler's simile, moving along or with the object in order to understand the object. Like the artist, the STM user also becomes a producer of images. The STM user becomes incorporated within the STM apparatus, and involved in the image-making and processing operations. In engaging in haptic vision, viewers of such images also become producers; like the artist, viewers become more apt to trace over what the images show than to stand back at a distance, like the art historian.

Like the artist's or art historian's experiential background, training or habituated response becomes part of seeing the images that the STM produces, as well as using the microscope—and so produces different capacities for response and understanding than with other kinds of vision practices. The capacities for response develop over time and with habit. As Nicolas Rasmussen explains in the case of electron microscopes, users spend great quantities of time with the instrument:

> Through daily practice learning about and working in the diminutive world made present by the microscope, the early electron microscopists underwent changes. As they familiarized themselves with its controls, the instrument became decreasingly refractive to their wishes, and increasingly predictable in performance. (229)

Like electron microscopists, STM users acclimate to the microscope, developing a habit of interacting with what they see. Acquiring a different habit of seeing and interacting also then expands a person's capacities for perceiving, altering her or his interactions in the world, especially as a person engages in vision practices not isolated to one instrument. Due to operating dynamics that are shared with other visualization technologies, the practice of using a particular visualization technology like the STM, and building up a particular form of visual practice that diverges from the dominant one of opticism, can also affect how the observer observes at other times, reinforcing haptic vision experiences in other contexts.

The observing subject practicing haptic vision is one who has cultivated a practice of engaging with the STM apparatus as well as with the imaged sample through observing the sample closely, either by moving slowly over the image to understand the information the image expresses, or by peering closely in order to respond to and anticipate manipulation. In the relationship created in the practice of haptic vision, the observer also changes. As haptic vision solicits not contemplation, but action and anticipation of future movements, the practice of haptic vision leads towards a subject habituated to haptic responses when confronted with similar images. Habituated response, then, builds on the bodily experiences of interaction to solidify the haptic habits the STM apparatus encourages.

The object of knowledge, such as the atom, is also altered through interaction with the subject accustomed to such haptic habits. As Taussig explains, "Habit offers a profound example of tactile knowing . . . because only at the depth of habit is radical change effected, where unconscious strata of culture are built into social routines as bodily disposition" (25). Therefore, because the habituated practices of haptic vision lead to a different position for the observer—whether that observer is an STM user or viewer of the images that the STM produces—and because vision becomes possible through movement and touch, scientific objects such as the atom become visible differently from how the objects would be visualized within a different vision practice. Grappling with atoms becomes a reality, even though it is a reality that occurs only in special circumstances, such as four degrees Kelvin, the temperature at which the xenon atoms were stilled enough for Eigler and Schweizer to position them in the "IBM" image production. The scientific objects that become visible with the STM and attendant vision practices, then (such as the particular understanding of atoms I recounted in Chapter 1), become possible given a certain kind of vision in certain conditions, conditions that help to form tangible and manipulable atoms, and in which certain acts and images become more persuasive than others.

STM-Haptic Vision and Rhetorics: Sketches of Implications

Vision practices frame the possibilities of understanding or reading images like STM productions, and through them, objects of scientific knowledge. For example, following only perspectival vision practices in the "Quantum Corral" image (Figure 4), as I further discuss in Chapter 3 and Chapter 4, produces a landscape picture that, while engaging in haptic vision practices, allows another reading of the image as datapoints and an understanding of the cone structure of atoms as the path of the tip as it moves in response to the surface atoms. The haptic vision practices described in this chapter operate amidst interactions among the STM user, the microscope apparatus, the imaged sample, and the viewer of the images produced by haptic interactions. Haptic vision practices have implications for rhetorics: Emerging out of haptic interactions are vision practices that re-conceptualize what is seen and reconfigure the relation between such standard elements of a rhetorical situation such as speaker/writer/imager and audience, subject, and object.

The observer's engagement with the STM suggests that what may be persuasive in the practice of haptic vision may differ from what may be persuasive when a viewer or user engages in other vision practices such as perspectival vision—and, returning to Lakoff and Johnson's point about bodily metaphors mentioned in this chapter's introduction, what may be persuasive to haptic vision practitioners may affect more than images. What may be persuasive while practicing haptic vision may signal that the reconfiguration of elements affects not only the interactions of our bodies with instruments, but also our conceptualizations of bodies of knowledge. Therefore, while practicing haptic vision trains observers to see differently, developing the habit of haptic vision trains viewers to observe the rest of the world differently.

What persuades practitioners of haptic vision seems to depend on the interactions viewers engage in and through with images as objects of scientific knowledge, as in the case of the STM. For example, the raised dots on Eigler and Schweizer's "IBM" images solicit observers to interact by forming letters and meaning from the spatial arrangements of light and dark spots present on the images. What is persuasive about this solicitation? My account of haptic vision practices suggests that what is persuasive about the interaction, as implied by the "IBM" images, is partly the haptic quality of the interaction. Based on the habit of running one's eyes over the dots, the eyes become similar to fingers reading Braille, and move the viewer—literally as well as figuratively. In so moving, the image persuades. In the "Quantum Corral" image, colors and peaks leap out, and we fall into the tactile and colorful waves of the image, suspending meaning as we first experience what the image expresses. The haptic qualities of STM images contribute to procedural and participatory rhetorics, particularly through the inclusion of affect as an inherent part the communication of the image: Bodily affect becomes a crucial part of the communication process. Additionally, the images are suggestive in how they present an argument based on the haptic practices of feeling the bumps *and* the more optical process of reading the bumps as letters, an argument that includes both significatory and asignificatory elements. (See Chapter 3 for further discussion of such combinations.)

Haptic vision helps make visible not only concepts of a tangible, atomic world and atoms, but also helps make the observer possible, one who can envision atoms through enhancing her or his interactive capabilities. What becomes visible becomes so through moments of blindness, the observer becoming immersed and reacting not though conscious sight, but through bodily response within a larger pattern of response. The haptic vision practices I have charted in this chapter, while still coexisting with the dominant, perspectival scopic regime as well as with others (such as those sketched here), do not only differ from perspectival vision practices. Through mimetic dynamics, haptic vision practices also invite a different, transformative relation between viewer and viewed, subject and object. New relations between observer and observed suggest a re-organization of the senses into a synesthetic vision practice and different interactions with the scientific object, such as the atom. New relations between observer and observed and subject and object occasioned by haptic vision practices also lead to different rhetorics and different formations of science and scientific knowledge through rhetorics and scientific objects. The next chapters turn to how the practice of haptic vision, along with the emphasis on interactivity, functions within rhetorics that help form nanotechnology.

3 Haptical Consistency: Emerging Conventions of the STM Image-Interface

In his 2003 *Nature* commentary, "Is a Picture Worth 1,000 Words? Exciting New Illustration Technologies Should Be Used With Care," chemical engineer Julio M. Ottino discusses the excitement generated by scientific images, particularly scientific illustrations. Ottino argues that because scientific images can create excitement and persuade viewers that the possibilities depicted are now or will soon be reality, illustrators should take special care that their work is scientifically accurate (476). Among other inaccurate illustrations, Ottino cites the November 9, 2001 *Science* cover that depicts gold wires draped over carbon nanotube molecules (Figure 5). One feature of the *Science* cover image that particularly disturbs Ottino is the smoothness of the gold, as opposed to the details of the carbon. He comments, "This picture purports to offer scientific insight, but if the carbon atoms are visible, then the much larger gold atoms in the structure should also be on view" (476). Following dominant visualization conventions that use linear perspective to represent objects, where objects are depicted in consistent relation to each other, both gold and carbon atoms should be visible in the same way; therefore, following Ottino's implication, the *Science* cover image misleads its viewers because the atoms are not consistently expressed.

However, Ottino's seemingly solid basis for critique wobbles on an examination of the wider context of nanoscale imaging, including other images of the nanoscale that illustrators could consult to depict the wires and nanotubes. As Chapter 1 elaborated, current technologies for visualizing the nanoscale, such as the STM and its close relative, the atomic force microscope (AFM), create images of electronic properties or topography via non-optical, non-lens-based methods; nanoscale images such as those created by the STM do not necessarily present information about visual characteristics as depicted by photographs, for example. While *Science* does not mention the production process of the cover illustration, the article the image refers to includes an image made with the AFM, developed to visualize non-conducting surfaces by dragging a probe just above a surface, thus interacting with surface electrons through proximity as opposed to the conductivity used by the STM. The study described in the *Science* article uses the AFM like the STM to manipulate—in this case, to select and move nanotubes onto the gate wires (Bechtold et al. 1317–1320). It is possible, even likely, that the illustrator or scientists commissioning the *Science* cover illustration used the article's images as references.[64] To return to the image Ottino critiques, if electronic properties are showcased, then gold atoms may end up "looking" quite different than other atoms with different electronic properties. Therefore, there may be a valid reason for why gold and carbon may not look similar in this illustration; furthermore, portraying them differently may actually express the scientific insight asked for by Ottino.

These two possible, reasonable and yet divergent, readings of this image illustrate a current shift in scientific communication away from more traditional image-viewing expectations and conventions, and towards expectations and conventions occasioned by the function of images to communicate data—and by the attendant vision practices viewers engage in as they create and view images of the nanoscale made by technologies such as the STM. Ottino, a chemical engineer and professor at Northwestern University who also designs journal covers, including covers for the American Institute of Chemical Engineers (AIChE), is hardly an idle or uninformed commentator (Ottino, "Role"). The fact that *Nature* included the title of Ottino's commentary on *its* cover suggests that Ottino's is not a solitary or marginal opinion. Indeed, as Ottino's criticism of the cover of one of the leading English-language, general science journals in another journal of the same caliber and similar readership attests, Ottino's comments and the *Science* cover image circulate in broader discussions about scientific visualizations taking place, as scientists and others grapple with the effects of using digital images to communicate non-visual, scientific data.

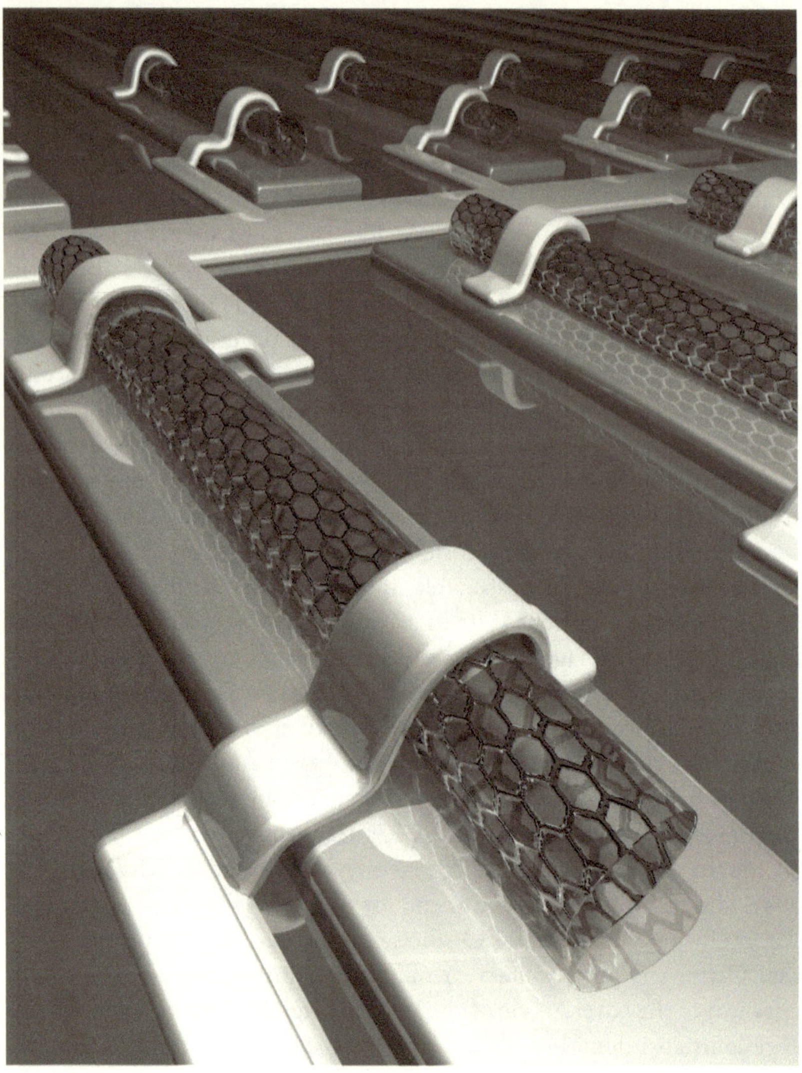

Figure 5. Journal cover depicting carbon nanotube gate wires. From *Science* 294. 5545 (9 Nov. 2001) (cover illustration). Reprinted with permission from AAAS.

Paying attention to discussions about scientific visualizations is important for understanding scientific communication. Ottino's reading indicates one way in which the increase in using images to display data might currently complicate understanding what images produced by, or even influenced by, visualization technologies like the STM convey, thus highlighting the importance of articulating any shifts and their possible effects. Peter Galison has argued that electronic images in physics can function as a common or trading language, a wordless creole (*Image and Logic* 52). Following Galison's comment, in nanotechnology, digital informational images become important, as they help to communicate across the disciplines forming nanotechnology.[65] Indeed, as Martin Rudwick demonstrates in the case of geology, a key part of the formation of a new field is the development of a visual language for the subjects and theories of that field (177). Further, scientific images of models or diagrams take on more weight in developing fields because images can create viewers' understanding of the science communicated, and so influence developing scientific theories and contribute to the directions of research programs.[66] As I mentioned in the introduction to this book, even the definition of nanotechnology is still being formed by scientists, granting agencies, the government, and in the popular imagination; images like those produced by the STM contribute to the definition of nanotechnology by making certain possibilities visible, thereby creating visions of what nanotechnology might be. In turn, the visibility of some possibilities (and not others) influence directions for nanotechnology funding, and influence how various groups engage with what technology is developed. Therefore, divergent understandings of what images convey affect discourses of and about nanotechnology.

This chapter explores some of the effects of using digital images such as those created by the AFM and the STM to communicate non-visual, scientific data about nanotechnology. More specifically, I analyze the use of visual conventions through a focus on the digital images made by the STM as scientific image-interfaces to express data about the nanoscale.[67] As Chapter 1 charts, the intensified interactions that STM users and STM image-viewers engage in as they focus on data produced by interaction with the nanoscale, coupled with the affordances of the STM apparatus, allow the image to function as an interface for interaction with the nanoscale and with data. Expansion of STM image functions to include interacting with atoms and with data contributes to how digital images like those produced by the STM blur boundaries between pictures and data as well as still images and interfaces. STM images, for example, may look like pictorial representations of objects, or may seem to be fixed images; yet, the images may shift—to database, or to interface. STM images may shift on the screen or in print, immersing the user within the practices of haptic vision discussed in Chapter 2. This chapter extends the work of the first two chapters, as I analyze the ways in which rhetorical conventions may shift in response to the deeply interactive and haptic production and viewing practices associated with the STM. (There are certainly other rhetorical elements present in STM images; the next chapter examines the appearance and possible modifications of two tropes common to scientific discourse also found in STM images.) While exploring the ways in which some widely-circulated STM images express a shift in visual conventions and expectations, I argue that the shift in form, combined with the focus on depicting data, creates different possibilities for persuasive expression that affects the conventions used in visual inscriptions of the nanoscale. Following effects on conventions becomes significant for rhetorical analysis not only because the expanded functions of STM images alter conventions, but also because informational images enlarge the possibilities for rhetorical production and analysis, as I explain below. Before turning to the conventions of STM images, I first demonstrate the rhetorical value of conventions in the process of understanding discursive change in order to clarify the significance of conventions in STM images.

Visual Conventions and Discursive Change

> Right, and so you pick something up out of an image that you wouldn't—you wouldn't be able to capture that same information in words easily, and so by having some key features in an image and highlighting those features, you can really pass information in a very direct way. I mean there's a kind of rule of thumb that when people are thinking about reading a paper, they look at the abstract and the figures and then they might look at the conclusions and then they decide if they're going to read the paper or not. [68]

This statement about reading STM images by a scientist who uses the STM includes a few general expectations about scientific images created for publication that inform how scientists create and read visualizations. The expectations about images meant to present evidence,[69] found in this scientist's statement and in the conventions that help scientific images meet his expectations, are rooted in the traditions of scientific image-making. Michael Lynch explains the importance of conventions for representing scientific knowledge: the conventions "take their authority from previous experience and the state of the scientific field to competently build on a body of assumptions about the represented structures" ("Externalized Retina" 163). The general scientific conventions that create and are in part created by the expectations mentioned above provide a context for understanding the role of conventions in communicating scientific knowledge so that viewers can "pick something up out of an image."

The scientist's expectation that images convey information not easily expressed in words is common to both scientists and non-scientists. In the case of imaging data, the expectation of communicating data through image form is based on the ability of humans to make detailed visual comparisons. Edward Tufte cites our ability to visually compare as the reason for our tendency to spatially organize great quantities of information presented in images (168).[70] In a widely-cited study of scientific representations, Lynch observes that scientific visualizations focus on selecting and mathematizing data for depiction in images, using such techniques as filtering out background; creating uniformity; making borders, shapes, and divisions clearer; and defining to show differences ("Externalized Retina" 161). Data selection techniques help to make scientific data a good fit for the image form by allowing for detailed visual comparisons. The organization of scientific data into an image also allows the informational image to appear as a coherent whole, and allows for associations with other kinds of images.

The expectation that one can not only create an image with "some key features," but also highlight those features, is in keeping with established scientific conventions of visual evidence production, where images designed for publication exhibit a few regular organizational patterns that also carry out the functions Lynch describes. Klaus Amann and Karin Knorr Cetina provide a description of three analytic orderings in their ethnography of how scientists produce visual evidence that offers a relevant general frame for understanding how the key features that scientists expect to see may be highlighted. Amann and Knorr Cetina's first analytic category is that of three-dimensional, perspectival organization used to foreground some features and relegate others to the background (112). Amann and Knorr Cetina's second category, presentation of the feature "within a matrix of other signals" (114) or features, provides a context for the highlighted feature. One reason for this context, as science studies scholar Françoise Bastide discusses, is that in a series, meaning is drawn from comparing images within the series to find significant differences. While comparison can be explicit, Bastide observes that the comparison can also be implicit: The viewer may need to know about other images relevant to the image under view (197). Bastide also observes that sometimes explicit or implicit comparison is not enough: "Often it is also necessary to have a key—or, rather, a well-established habit of thought—to be able to interpret a photograph" or other image (197).[71] The need for a key or pointer relates to Amann and Knorr Cetina's third analytic ordering, reliance on "pointers"—marks added to an image to indicate some features as significant (114). "Pointers" can be textual marks, arrows, or other marks to direct the viewer's eyes to the important data. These three common ways of organizing data form general conventions that align with scientists' expectations for creating and reading scientific visualizations as inscriptions.

The expectation expressed above that an image can "pass information in a direct way" also depends on the very common assumption that the visual form of the image inherently contains a less mediated, more transparent way of communicating than text, for example. However, expectations of direct transmission of information do not reflect how images function in scientific texts; as Amann and Knorr Cetina explain, images do not add to scientific discourse by "displaying the data unequivocally, by adding the certainty of proof which the text can only refer to, but not 'show.' . . . Images, perhaps more than texts, provide infinite opportunities for visual exegesis, thereby functioning to keep the discussion open, not closed" (115). How the contradiction between the expectation of direct communication via images and their actual, more open-ended communicative function plays into misunderstandings and discursive disjunctions is one reason for further analysis of images within the contexts of the discourses in which they are formed and contribute to order to better understand how images do convince potential readers of the significance and worth of images.

One way to explore how visualizations communicate within the context of a community—how images help convince viewers to read a scientific paper, for example—is through how image-makers use visual conventions that allow viewers to understand the significance of what the image shows. In part, visual conventions help meet viewers' expectations and communicate data persuasively through acting as an exordium to predispose potential readers of an article to engage with its argument, similar to how Teena A. M. Carnegie argues that interfaces work rhetorically. However, visual conventions function rhetorically in other ways as well. A closer look at how visual conventions function helps not only to identify them, but also to explain the significance of focusing on conventions as a unit of analysis.

Comparing the analysis of visual conventions with another main way of assessing the rhetorical impact of images reveals some of the advantages of analyzing at the level of conventions. While some rhetoricians productively explore the organization of visual elements through gestalt and other cognitive principles,[72] many rhetoricians theorize how images are communicative, persuasive acts that convey scientific understanding through a comparison of images to verbal language. Rhetorician of science Alan Gross argues for the rhetorical value of visual images in science through an expanded understanding of "language," creating parallels between visual language and scientists' use of other "artificial" (as opposed to "natural") languages such as logic or mathematics to communicate through forms such as images and equations ("Darwin's Diagram" 52). Therefore, if scientists use artificial language to communicate through images, images can be "read" only if one understands the elements of that language. Emphasis on the purposeful construction of images created by parallels with verbal language helps counter many viewers' assumptions that an image presents an unmediated, "natural" view of reality. A focus on the "visual language" of images thus encourages a more critical perception of an image as constructed communication. The focus on language, as Rudwick explains, also emphasizes that a discipline or community's visual language must be learned, and may change over time (151). Other rhetoricians compare the elements of an image to language in order to explore its visual rhetoric, or focus on identifying a semiotic grammar of visual images.[73]

However, analogies between visual and textual language reach a limit due to differences related to function and form. Ellen Lupton and J. Abbott Miller critique the term "visual language" as leading towards isolating visual from other forms of meaning-making (65). Charles Kostelnick and Michael Hassett identify a problem with only drawing from analogies between verbal and visual language in order to understand how images communicate: The analogies "don't supply a macrolevel structure for understanding how it [visual language] functions as a language" (1). While comparison with verbal language can produce insight and provide a sense of order, the order inherent in a visual communication may or may not have much in common with verbal language. Following Susan Oyama's argument that form and information develop ontogenetically, forms have unique, material properties that influence how data is expressed within them; thus, forms influence how what is communicated is expressed. Focusing on conventions can move beyond the limits of the comparison with language, and yet still emphasize the intentional nature of image composition as well as the form-specific and function-specific aspects of the image. Focusing on convention does not negate the possibility of the existence of a visual language, nor does it suggest that conventions do not participate within a larger structure. Instead, focusing on convention sets aside a sense of implied structure to allow for attention on how form is created while still accounting for the constructed nature of images and the use of images as communication within a discourse community.

Another reason that the comparison of visual communication to language is limited has to do with how the form of the image affects the communicative function of the image, especially because different forms may include different, multiple functions. Scholars have argued that scientific visualizations can contain propositions, or reasons for an argument that express a truth value, but can also function in non-propositional ways (Blair 44).[74] James Elkins summarizes the finding that scientific images function in multiple ways: "sometimes they illustrate or propose theories; other times they merely *exist*, taking a certain place in the chain of written discourse and modifying it in ways that are difficult to describe" (*Domain* 39). Focusing on conventions forms a way of enlarging the scope of analysis beyond the elements of language to include visual components that might not map onto a grammar, for example. J. Anthony Blair explains that all arguments need not be propositional ones in defending the possibility that arguments can be visual (48–49). Therefore, a focus on both propositional and non-propositional elements for how they "modify the chain of discourse" may be relevant to the rhetorical impact of visualizations.

Analysis of conventions also highlights the influence of conventions on collective reading practices—something particularly important for scientific inscriptions given that, as Elkins explains, the meanings of what Elkins calls "non-art" images change more quickly than art images "because they depend less on resemblance and more on specialized interpretative skills that are easily lost over time" (*Domain* 36). Elkins cites two examples in which scholars trace how interpretations of scientific visualizations change over time. (Elkins cites David Kaiser's history of interpretations of Feynman diagrams and Michael Ruse's account of

Sewall Wright's use of landscape-like diagrams to visualize population growth.) Elkins attributes some of the change to the strength of pictorial conventions that were common to a wider audience than first viewers of the images. He claims,

> it is the specific appeal to pictorial conventions that ensured both that the images would be influential and that their influence would be partly unpredictable . . . . Feynman's diagrams and Wright's maps changed meaning because people were drawn, inevitably, to use them as if they were more than superficially related to the common kinds of pictures they both resemble. (*Domain* 38)

Elkins's insight shows the importance of analyzing visual inscriptions at the level of conventions to explain not only the pull of particular images, but also the ease of adaptation of these images. Unlike a text, images have tendencies to slip their semantic moorings. Focusing on conventions reveals one level of the communicative dynamics of images.

Following conventions in scientific visualizations is also useful because a scientific image often contains different forms or modes of presentation at the same time. Often, different features of the image depict different properties or kinds of data, what Elkins calls "multiple referential modes" within the same scientific image that help produce viewers' understandings (and misunderstandings) (*Domain* 36). As Elkins explains, while other images (such as fine-arts images) may contain more than one mode, but highlight one mode as dominant, information-rich images tend to "blend more widely disparate modes" (*Domain* 36). Different modes of data, then, complicate reading and analyzing scientific images like those produced by the STM. For example, "Molecular Switches" (Figure 6), an image made from STM data, features two perspective views of the same sample that show differences after the experimenters added molecules to another layer of molecules. On the flat background, diagrams of the operating switch also appear. To read such images, Elkins suggests, viewers must switch between these modes—if, of course, viewers are familiar with the requisite reading practices (*Domain* 36).

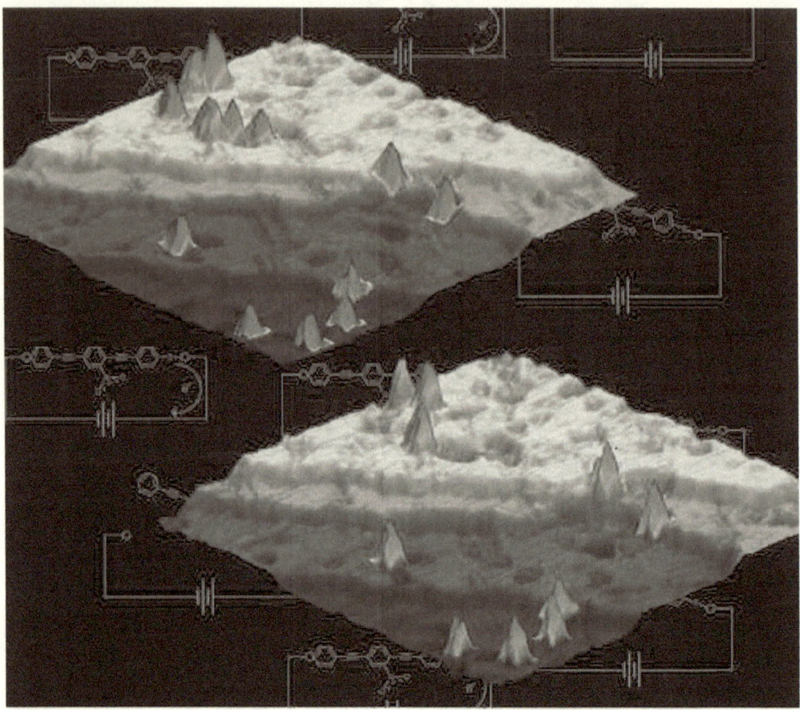

Figure 6. "Molecular Switches" (Donhauser and Mantooth). Reprinted with permission from Dr. Paul Weiss, Dr. Zachary Donhauser, and Dr. Brent Mantooth.

Conventions are an inherent part of viewing practices. As Kostelnick and Hassett explain in describing the characteristics of visual convention, conventions first need to "solve information design problems that many users share—*typical* problems, not novel or unique ones" (79). For example, in the case of STM imaging, how scientists depict electronic properties becomes a typical problem due to the affordances of the microscope. Kostelnick and Hassett also observe that conventions must be seen regularly by those within the group in which they emerge; conventions must be "reasonably economical to imitate," and those creating the images must be persuaded to use them instead of other conventions (79). The *Science* cover that Ottino critiques presents an interesting case in point, for it seems that the illustrator uses depiction techniques that are conventionally used in AFM images in the *Science* cover illustration, and so the depiction of electronic properties in the image become persuasive enough to migrate to a different kind of image.

Conventions also suggest the community or cultural reading cues provided by the image producer without suggesting that the individual producer's intention or aesthetic choice is the only driving factor in image production. As conventions exist at a collective level, conventions are not invented or replaced on a whim; instead, they are replaced only when users must do so. As Kostelnick and Hassett explain, imperfect conventions remain even when a better replacement is available due to readers' habits of use (79). Therefore, altered conventions mark significant shifts in reading and viewing habits, indicating that either the previous conventions stopped functioning entirely, or that readers' use habits have changed.

The characteristic of changing conventions only when necessary also identifies the possible scope of sources of influence on changing conventions, including factors such as new knowledge and technology, as image creators envision information in different ways due to the affordances of new technologies or ideas, Kostelnick and Hassett explain (43). Focusing on conventions emphasizes not only the choices of individual image creators, but also the choices available to them. In other words, a focus on conventions emphasizes the cultural practices from which the individual draws in order to communicate to the audience. Those who create STM images, for example, must rely on shared visual conventions to persuasively convey experience with—and evidence of—the nanoscale.

Focusing on conventions also highlights the rhetorical dynamics of discursive change. As Kostelnick and Hassett explain, those creating images often alter conventions as image creators adapt images to particular viewers in a specific situation (6). A focus changing conventions, then, may also indicate the significance of a change in viewers or situation. For example, asking whether alterations occur when scientists adapt to creating inscriptions that also function as interfaces leads to understanding the significance of the interactions recounted in Chapter 1. Exploring whether alterations occur when STM image creators make visible atomic phenomena that are unrepresentable according to quantum mechanics can indicate the significance of STM visualizations. Elkins identifies one dynamic of conventions relevant here in a comment made within a discussion of the limits of representation, particularly with regard to scientific images. He writes, "As in the history of art, images of unrepresentable objects put a strain on the pictorial conventions they inherit, finally breaking them and becoming different kinds of pictures" (*Domain* 44). Elkins's comment suggests that a focus on convention is also important in how convention analysis can chart differences in the "kinds of pictures": How significant is the shift from "image" to "image-interface," for example? A focus on conventions in STM images forms one way to explore questions about discursive change and its possible significance.

Reading Conventionally: Linear Perspective in STM Images

At first glance, many STM images, such as the "Quantum Corral" image (see Figure 4) published in *Science*, as well as a color version published on the cover of the journal, look like conventional representations of objects—like rings of marbles or spikes in the color version, for example. Even seeing multiple images made from the same data leads to a feeling of recognition of objects. Other STM images also create the impression of familiarity, such as the images showing D. M. Eigler and E. K. Schwiezer's manipulation of xenon atoms (see Figure 1 and Figure 2). Even the first processed image from the STM made by its inventors presents familiar, recognizable aspects (Binnig and Rohrer, "From Birth" 400). The very proliferation of STM images that look like representations of objects provides a justification for Ottino's interpretation with which this chapter began, indicating the scientific convention of showing three-dimensional data in a two-dimensional space. While the apparent use of the conventions of linear perspective to fit within scientific convention that nanoscale images like Figure 1 and Figure 2 exhibit is not the only way that data from the STM has been presented, use of the conventions of linear perspective commonly appear in STM images that present nanoscale phenomena in various contexts, including scientific journals, the websites of scientific groups, government documents, and more popular and educational web sites. A closer look at the traditional convention of linear perspective as the convention appears in images reveals some of the rhetorical force, as well as the limitations, of the convention.

The use of linear perspective serves as a visual organizing device through the generation of single-point perspective and recognizable shapes or objects. In a study of visual persuasion in early modern painting and architecture, Caroline van Eck claims that because perspective directs the gaze by establishing a single point of view with a vanishing point, and so constructs a "visual hierarchy," the painter's role includes choosing "a disposition that is most conducive to persuasion" (25–26). The painter's use of perspective as a framing device is associated with visual practices that have formed a dominant, almost ubiquitous, strain in Western art and images since the fifteenth century (Alpers xx). The dominance of perspective suggests that the conventions of linear perspective reach the level of background norm for viewers: As Elkins remarks, "conventions of computer-generated perspectival scenes in military and scientific simulations, architecture, and commercial games appear 'natural' or mathematically driven to their designers, even though they can be shown to derive from Western landscape painting of the last two centuries" (*Domain* 9).

Some of the key characteristics of linear perspective explain the traditions of conventions that guide readings of images; the conventional characteristics of linear perspective also explain some of the rhetorical functions of these images. Scholars such as Martin Jay have labeled the set of visual practices associated with perspective "Cartesian perspectivalism." The view of perspective that Jay discusses emerges from the writings of the tenth-century Alexandrian optician and mathematician Ibn al-Haytham (also called Alhazen) (Nelson 5). Svetlana Alpers traces the beginning of the dominance of Cartesian perspectivalism in European art from the fifteenth to the nineteenth century, back to Leon-Battista Alberti's description of perspectival drawing technique in his *De picture* (Alpers 22).[75] Van Eck observes that one of Alberti's innovations was to develop a theory of art from the viewer's point of view, allowing for the analysis of a painting's effects on viewers (26). Perspective has remained the main Western mode of seeing the world and making images, so much so that linear perspective has been called a dominant, "scopic regime."[76]

Many writers attribute much more than pictorial practices to the dominant visual practice of perspective, particularly in regard to science. For example, perspectival characteristics became associated not only with Italian painting in the Renaissance, but also through Johannes Kepler's differentiation of light rays and perception—the focus on the former with science (Illich, "Scopic" 3). Ivan Illich and others argue that popularization of the Albertian model of perspective actually marks the transition point to a conception of the eye as instrument, which others attribute a little later to the introduction of the lens in the microscope and the telescope (Illich, "Guarding" 51). As discussed in Chapter 2, William Ivins argues that an exactly repeatable pictorial statement formed one of the main tools of the scientific revolution of the Renaissance. Ivins claims that the ability of perspectival techniques to create consistent or regular images formed a crucial point in the development of science and technology, because the consistent repeatability of a pictorial statement makes it possible for scientists (and others) to understand how a representation of an object corresponds to an actual one, and to recognize objects they have seen in images (17). Ivins's observation was picked up by science studies scholars such as Bruno Latour in his also-influential formulation of the way that visual inscriptions become persuasive in scientific knowledge production. Latour argues that "optical consistency," as Ivins terms it, becomes "a regular avenue through space," and allows inscriptions to become mobile (i.e., as able to travel to other contexts) and immutable (i.e., as able to remain consistent, even though inscriptions appear in different contexts) ("Drawing Things Together" 27). Friedrich Kittler suggests that perspectival representation is more than just an artistic or scientific form. He argues that perspective and its emphasis on place value—found in the popularization of Alberti's perspectival painting techniques and in Gutenberg's type layout techniques, both of which allowed images and text to be printed and reproduced—was a major source of European power (Kittler 46). These arguments suggest that linear perspective is more than a drawing technique: Linear perspective has become a dominant, conventional way of perceiving and depicting the world. The dominance of the conventions of perspective adds the impression of authenticity or reality to the use of linear perspective in images.

Pictorial techniques associated with linear perspective, such as single-point perspective for the viewer, and shadowing to indicate the three-dimensional status of objects, function as visual conventions in STM images. For example, the black and white quantum corral images published in M. F. Crommie, C. P. Lutz, and Eigler exhibit certain qualities associated with linear perspective that we use to recognize objects: a circle of what look like marbles, within which exist concentric waves or ridges centered by a peak (219). The play of light and shadow indicates perspective; the circle being centered in the image suggests a focus on the circle as an object. Familiar elements also suggest that reading a corral image, for example, as depicting an object at a distance is appropriate as well. The corral image published on the cover of *Science* (Figure 4) also fits neatly into the overall impression of a perspectival representation of objects through the marking of shadow and a shift in the angle from which the observer sees the corral. Even Gerd Binnig and Heinrich Rohrer's presentation of the first STM image as an object, rotated to show perspective, suggests that the image presents a view of a three-dimensional object ("From Birth" 400).[77] Perspectival details create the sense of an object that aligns with the Albertian model's view of a representation: "an optical substitute for the object itself. . . . It is a facsimile emitting the same bundles of light rays that would be reflected by the object if it were there, beyond the picture's frame" (Illich "Scopic" 12). Reading STM images wholly within the context of perspective creates an impression that STM images signify objects within perspectival space as "objects of mastery" to the viewer, as Colin Milburn argues (*Nanovisions* 73).

Perspectival conventions function rhetorically in these images to persuade viewers that STM images present depictions of solid, believable objects. However, while the use of linear perspective thus functions as an immediate appeal to credibility, the use of the convention also performs the work of arrangement in that data are shaped into what seem like familiar, approachable objects, lending a sense of authenticity to the image. The use of perspectival conventions also establishes a common ground with the viewer, as van Eck argues happens in early modern painting (55-88). Establishing a common ground allows the viewer to feel as if he or she is immersed in a physical place (the nanoscale) among physical objects (atoms and molecules) so that the viewer identifies with the researcher, from whose viewpoint the viewer sees. Additionally, the presence of linear perspective can also create identification with what is represented. As van Eck notes, the main point of visual persuasion, according to Alberti, and found in early modern painting, is "the identification of the viewer with what is represented, resulting in a sense of living presence" (56). The long-standing convention of perspective, then, has helped train viewers of images using linear perspective to look for a sense of presence and identification with what is represented. Steven Shaviro's description of how images affect viewers before they have time to understand their meaning further helps to explain the effectiveness of perspectival conventions and the consequences for understanding what they show, even if we "know differently" (26). The techniques of linear perspective have already organized the viewer-viewed relation in terms of distanced observer and observed object. What becomes visible for viewers who do not have experience with quantum mechanics is the appearance of atoms as objects composed in familiar ways—so familiar that these objects can even spell out letters, as in the "IBM" image. The use of perspectival conventions thus helps image creators argue for the credibility of the STM, the credibility of the image creators' work to other scientists, and also for the existence of the nanoscale as well as the stability of atomic phenomena as research objects. The dominance of linear perspective conventions in Western visual culture contribute to the rhetorics of STM images by making perspectival organization of data seem authoritative, as less subject to questioning.

However, perspectival conventions do not provide a completely satisfactory account of what and how STM images communicate. While the use of linear perspective conventions in STM images may persuade readers that nanoscale phenomena actually exist by creating similarities with objects on the human scale, the links between nanoscale and human scale can also obscure some of the very qualities of the nanoscale that have intrigued researchers.[78] As Ottino observed about the cover image he critiqued, depictions of the nanoscale in STM images, particularly in regard to how atoms are made visible, does not fully match the concept of atoms in quantum mechanics—a concept used in constructing the STM apparatus, where electrons are articulated in terms of wave and particle, of constant movement and exchange with other atoms, and where atoms are not individual objects with defined boundaries. Yet, as Chapter 1 demonstrates, some of the affordances of the STM create an experience of atoms as tangible and bounded; the representational conventions that cause atoms to appear as objects add to understanding STM images as representing individual objects.

The conventions of linear perspective help to shape interpretation, yet perspectival conventions do not always only help scientists communicate the data due to the very dominance of the conventions. STM images exhibit the sway of convention, for example, in the case of the corral images. As Eigler explains, the main features he wishes to demonstrate in these images are the waves, not the seemingly solid circle of atoms (although the atoms also convey impressive atomic manipulation):

> The symmetry and simplicity of the waves in the interior of the corral have a special meaning to a physicist. These things were entirely predictable, and we routinely work them out in our problem sets when we take courses in quantum mechanics. We knew of them in a purely cerebral way. But here they are, alive to our eyes. (Frankel 261)

Due to the perspectival focus on objects within a space, viewers may not regard the space "between objects" as an important component of what the image producer aims to communicate. That the waves in the corral images are not as emphasized for non-physicist viewers is apparent in the introductory note that appears in an *American Scientist* article by artist Felice Frankel on the quantum corral images: "The reader should take note that an important visual component of both images is the representation of the electron eigenstates—wavelike structures—within the corrals. The drama of the atomic peaks might distract our eyes, but with such amazing science, does it really matter?" (261). While Frankel helps readers understand the importance of the waves, her last comment shows the tension between "the drama" of the objects (what we might think is depicted) and the "amazing science" of what the image "actually" shows. Here, too, then, the use of the linear perspective convention causes friction in communication.

Even if we discount that what we see in STM images are objects—if we know, for example, from quantum mechanics that atoms are not tangible, static objects, and tell ourselves that the image's formation of an object is only metaphoric (something that is difficult even for trained observers reading visualizations of non-visual data; for example, see Dumit 68–69)—perspectival conventions also configure a space where the observer stands at a fixed, distant point in order to view an object. What STM images depict does not exactly match the relation between fixed observer and distant object, for the space imaged through the data—that of the nanoscale—has no horizon, no perspectival vanishing point, and no point at which an observer can be extrapolated to stand. Another example, from Ottino's commentary in *Nature* about the *Science* cover image of the nanoscale mentioned above (Figure 5), shows the reliance on a view of a consistent space between the observer and the observed that perspective creates. Ottino objects to the use of "shadows, reflections and so on as they would occur in everyday life, which most definitely do not occur in quite the same way at the nanoscale" ("Is a Picture" 476). Implying that such reflections and shadows are merely inaccurate, Ottino suggests that it is indeed possible to envision this image as perspectival, just at a smaller scale. Ottino does not explicitly question the possibility of perspective at the nanoscale, or of representing nanoscale objects according to perspectival conventions. Instead, Ottino suggests that the illustrator merely did not apply shadows correctly. The fact that it is possible to misinterpret what images made with the STM (and the AFM) show as objects is at least partly enabled by the dominance of perspectival conventions in such images. We expect to read STM images as static images, as we would a painting. However, questions about STM images build: In what set of conventions could different atoms be depicted differently? How might atoms be depicted at all? How might even the "false" color in some images (such as Figure 4 and Figure 6) disrupt the seemingly smooth use of the perspectival convention? Such questions suggest further exploration of how STM images do communicate beyond the convention of linear perspective.

Emerging Conventions of Data Image-Interfaces

> *For the Science cover, in a sense, I started with an empty canvas (a black computer screen) and a concept about what I wanted to achieve. I began to apply paint. It was a matter of displaying the corral in 3-D and then searching for the combination of perspective, lighting, surface properties and color that communicated what I wanted to communicate. So the final image came to reveal itself in steps as I worked with computer knob-things.*
>
> —Don Eigler

On first read, this account of how Eigler created the color image of quantum corrals (Figure 4) seems to correspond closely with the process of painting a picture. Indeed, Eigler summons the traditional conventions of linear perspective in his description, particularly through the metaphor of painting and overt connections to specific techniques of "perspective, lighting, surface properties, and color" (261). Even the sense of working with "computer knob-things" could suggest that Eigler is merely painting with computers to form a visual work such as those early modern paintings van Eck describes. However, some of the more subtle details, such as Eigler's more specific words—that "the final image came to reveal itself in steps," for example—and even the recursive nature of "searching for the combination . . . that communicated what I wanted to communicate," indicate the interactive image-processing I describe in Chapter 1 (261). The image production processes Eigler discusses, and that other STM researchers engage in as scientists create images for journal covers as well as journal article figures, suggest that another look at how visual conventions such as linear perspective communicate data about the nanoscale. Additionally, a focus on form, following Oyama's observation about the ontological development with information of form, necessitates another look. How does the fact that STM images, like many other digital images made with recent scientific and medical visualization technologies, depict data and yet also function as interfaces, affect the use and rhetorical work of conventions, such as linear perspective, that appear within them? A closer look at form (the digital interface), and referent (data) leads to a reassessment of some of the techniques of depiction traditionally associated with linear perspective, and suggest the presence of an emerging convention of the data image-interface. The emerging convention of the data-interface contributes its own rhetorical pull and, as I argue below, needs to be read along with other conventions that are present (such as linear perspective), in order to understand the ways in which STM images function rhetorically.

One aspect of the form of STM images obscured by only reading the dominant convention of linear perspective is the fact that STM images function as interfaces as well as visual communications. The STM and the imaging processes afford researchers further interaction with STM images, so much so that the image becomes an interface, the site of interaction between researcher, viewer, data, and atomic phenomena. The interface function of STM images is in part enabled by the digital composition of STM images. The interface function influences the conventions expressed within the image, as STM image creators search for ways to solve information design problems.

The compositional attributes of the digital image form create affordances for STM image creators. Graphic designer and digital humanities scholar Johanna Drucker highlights the significance of the digital image form in relation to data, observing that "the graphic form of information, especially in digital environments, often *is* the information: design is functionality" (*Speclab* 17). If the design composes the data, and the data composes the design, then analyzing both together becomes critical to understand how a digital image works. Some other characteristics of digital images also become important for understanding the use of conventions. For example, Gilles Deleuze explains that with a digital image,

> The organization of space here loses its privileged directions, and first of all the privilege of the vertical which the position of the screen still displays, in favor of an omni-directional space which constantly varies its angles and co-ordinates, to exchange the vertical and the horizontal. And the screen itself, even if it keeps a vertical position by convention, no longer seems to refer to the human posture, like a window or a painting, but rather constitutes a table of information, an opaque surface on which are inscribed "data." (*Cinema 2* 265–66)

Following Deleuze, the organization of space in the digital image disrupts the reading of single-point perspective, and does not automatically create a space for a distanced observer or for a sense of orientation. Just as one can walk around a table and view its surface from different perspectives, one can vary the angle with which one views the inscription of data on a surface. The surface is also emphasized as opposed to the illusionistic space that linear perspective creates.

The relationship of viewers to images that the digital image articulates creates another characteristic of digital images: the appearances of the images themselves are variable, not only because the image creators can present data in different ways, but also because image viewers move through the data point by point, whether or not viewers can actually interact with the data on a computer screen. The alterability of the appearance of the image highlights the main function of all data images: to encourage viewers to read data rather than to see an object. The emphasis on reading data changes the information design problem from how to depict an object to how to depict an understanding of data; or, in the words of the scientist mentioned at the beginning of the section on conventions, to "highlight certain features" to convey information.

The form of the digital image thus fits well with the general scientific conventions of using images to depict data, especially in the case where the referent is already seen as data (as opposed to an object). Jonathan Crary's discussion of digital images aligns with the trend of expressing data in image form found in recent scientific and medical digital visualization: "if these images can be said to refer to anything, it is to millions of bits of electronic mathematical data" (2). The ability of the digital image to present much data, coupled with the conventions of scientific inscription to include multiple modes of data, create some interesting confluences, and may lead to different conventions used to solve information design problems.

Characteristics of digital images affect the ways readers approach the STM image as well. Returning to the quote beginning the section on conventions, scientific viewers of images created by the STM expect to see highlighted features of an image, and so employ different viewing practices to read the data than they would to view an early modern painting in an art gallery, for example. In addition, Lev Manovich's explanation of how readers approach new media images applies to digital visualizations such as STM images. He states that digital (new media) images "turn a viewer into an active user. As a result, the illusionistic image is no longer something a subject simply looks at, comparing it with memories of represented reality to judge its reality effect. The new media image is something the user actively *goes into*, zooming in or clicking on individual parts" (183). The image becomes an interface, and the viewer participates in the image through the interface as she or he compares data points. If the image is computer-generated, the viewer may even change the image as she or he zooms into sections of data, converts the image to a histogram, or manipulates the image to get a better sense of the data. Following differentiations in data, the viewer moves his or her eyes over the contours of the surface of the image; the viewer is not positioned as a distanced observer. The close reading provided by Figure 6, for example, illustrates the value differences between a group of blue pixels to those of a vivid red—observing that the value differences do not necessarily depict an object or even immediately reveal meaning—is important to understanding STM and other data images. As the viewer engages in haptic vision practices, he or she focuses on the surface of the image to read the data, not necessarily on possible "objects" shown within.

To enhance the viewer's reading of the data and highlight some features, STM image-makers use techniques that may at first be associated with linear perspective but that, on closer examination, prove to function differently. Techniques like the use of false color, emphasis on contour and texture, and use of time as a reading technique are not new;

they fit within the larger conventions of scientific imaging. However, techniques used to highlight features are used in conjunction with the image-interface form in such a way that the techniques suggest different reading practices and can also disrupt the reading practices associated with linear perspective. The techniques become part of the emerging conventions for informational image-interfaces.

Use of false color is one technique that indicates an emerging convention. While color appears as part of the traditional convention of linear perspective in that objects are often expected to be colored naturalistically in order to mimic reality, and so contribute to the sense of real-world objects that linear perspective evokes, color becomes a key indicator of something else in depicting data, as data has no inherent color. (Of course, color does not exist at the nanoscale; color is impossible to resolve at the level of atoms much smaller than half the wavelength of light, the minimum optical resolution.) The use of a color scale to differentiate data, for example, can be found in STM images in journal articles as well as on splashier journal covers. Black-and-white scales, or a sepia-and-white scale that is the default scale of some images produced by commercial STM imaging software, are common. Scales using other, more striking colors tend to appear on journal covers, group websites, or in other electronic or print publications designed for wider audiences. One scientist who creates images with the STM states,

> So yes, color does become very important in drawing your attention to certain points, and that's, again, where, like, the linear and nonlinear link, having certain little sections where you make only this very top that little yellow and you can control how quickly you go from red to yellow and how much—like what you call the top, where you saturate.[79]

The scientist adds, "for the most part, these images are colored by topography . . . . It's nice, it helps draw your eyes to the feature."[80] The use of color satisfies the conventional criteria that Kostelnick and Hassett describe in that color helps to solve a typical problem facing those creating STM images: how to guide readers' attention through the images (79). As Tufte observes, use of color is a common technique in differentiating information in images, though Tufte also argues that color is a difficult and complex technique to use (81). Because each image presents so much data, guiding readers' attention is especially important. The appearances of STM images on journal covers and articles, websites, and other promotional material help circulate the images that use color so that the color scales are regularly seen by audiences, thus forming one characteristic of a convention (Kostelnick and Hassett 79). Using color scales is also economical, as color scales are embedded within commonly accessible imaging programs, and the use of color in data images is familiar to both users and viewers. The fourth conventional criterion, that image creators must be persuaded to use a convention instead of other conventions, is manifested through the presence of color in a variety of images to show differences in data (Kostelnick and Hassett 79).

The use of color functions rhetorically to arrange the procedure of reading the image, as color visually guides readers through data. Color functions throughout readers' experience of data, for readers make connections between data points with similar colors. Color thus helps researchers argue for the significance of some data. Use of color in STM images is differentiated from use of color according to the conventions of linear perspective because of the arbitrariness of STM image colors. While color appeals to the audience's emotion, and functions to engage viewers in the image, the color scale used is not relevant to the meaning (or signification) of individual data points. What matters instead is that colors help readers follow how the image creator highlights the features of the data sample. The conventional shift signaled by color helps explain the lack of agreement about a common color scale in nanotechnology, as reported by Arie Rip and Martin Ruivenkamp, also explaining the comments from researchers I interviewed, who choose colors that appeal to them most (29).[81] Color does not function to verify authenticity or even participate in the rhetorical work of establishing credibility; instead, color functions as a reading device that performs the rhetorical work of arrangement while also appealing to readers' emotions and, at times, disrupting the rhetorical arrangement of the linear perspective found within the images.

Another technique that contributes to informational digital image communication is the use of contour and an emphasis on texture to communicate the significance of some data. Data differentiations are mapped in terms of distances from a baseline, creating a contour effect that often leads to an impression of a landscape as well as an impression of texture in detailed shapes. Use of the image as an organizing device partly creates a focus on contour and texture. Viewers can see the whole and also focus on specific sections (Tufte 31). What appear as peaks in the color corral image (Figure 4), are actually part of a contoured line that maps the differences in electron interactions between tip and sample. Tracing contours is a common way to arrange data so that the data is presented in relation to the other measurements. Use of distance or contour thus helps to solve typical problems of organizing and presenting large amounts of data in an image. As mentioned above, the focus on contour and texture is also a regular feature of many STM images, and is a feature of much imaging software. Contour and texture effects function with color to guide readers through the data by focusing attention not only on what stands out at the peaks and troughs, but also on the surface of the line and the material.

Another convention for differentiating data is through the use of time. Multiple images are presented either separately or in the same image, or in film—another variant of STM visualizations. Figure 6, "Molecular Switches," exemplifies the emphasis on reading for change in time. In "Molecular Switches," the fact that there are two "objects" of surfaces is not significant. Instead, what is important is the change from one to the other. The strategy of displaying multiple elements, what Tufte calls "small multiples," is useful in that readers need only train themselves to read one of the multiples for the data; designs presented similarly in the other multiples allow readers to focus on the data as opposed to figuring out the design (Tufte 29; also see Tufte, Chapter 4). The image series of assembling the letters I, B, and M (Figure 1) also points to a reading process that uses time in that the images show "new versions" of the image as the image producers manipulated the sample. The emphasis on change—in this case, over manually "refreshed" images—invites readers to move from image to image and, in some ways, go through the process of forming letters or the corral along with the experimenter as viewers read according to the scientific ordering convention of comparison. (See Chapter 4 for further discussion of these images and this process.) While the presentation of small multiples to show difference over time is not universal, the series communicates information about the dynamics of nanoscale phenomena and the images that can be collected from the STM.

Each of these techniques—use of false color, contour and texture, and emphasis on comparing images over time—helps form visual conventions for digital, informational image-interfaces. Viewers can use informational image conventions to steer through the vast amounts of data contained in an image-interface using color, contour, texture, and comparison between images in series as reading cues to respond to the data in the image and determine their next direction through the data. As viewers follow visual elements such as color, contour, and time to read the data, the visual conventions governing the use of elements functions rhetorically to arrange the data and, therefore, to function as a frame—not to frame an object within, but instead to frame the viewer's experience. The frame does not then impact the viewer as immediately as does a perspectival frame; an informational frame occurs in the act of reading, of experiencing the data. What does impact the viewer at first are the visceral, tactile effects of color, contour, and the play of changing pixels as readers move from one data point to another. The emerging conventions of informational images function through the image-interface as readers use haptic vision practices to read the data.

Reading Scientific Image-Interfaces: The Rhetorics of Conventional Dynamics in the Intraface

While reading the conventions of linear perspective and those associated with the data image-interface individually starts to create a sense of their rhetorical work within a STM image, Drucker's connection of design to functionality in digital environments suggests a closer focus on how conventions work in conjunction with the functions of STM images as scientific image-interfaces (*Speclab* 17). Analyzing the components of STM images in relation to how the images function as interfaces highlights the form-specific *and* function-specific conventions that help make STM images persuasive by articulating what elements become most important or significant, thus yielding further insight into how STM images communicate. Looking at why conventions related to uses of color, contour, and different explorations of time become useful enough to adopt, especially when these uses may disrupt established conventions, can explain the dynamics within discourses about scientific images that occasion critiques such as Ottino's. Kostelnick and Hassett's point that readers' habits of use often keep conventions in play, even when the conventions do not work perfectly, suggests a closer look at the elements that do not relate to the dominant conventions (such as linear perspective) used to assess the significance of changes that occasion change in conventions.

Communication and media studies scholar Alexander Galloway articulates the concept of the interface in a way that emphasizes the dynamics of an interface and the reading processes involved. An interface "is not a thing, an interface is always an effect. It is always a process or a translation" in that it forms a meeting of unlike things (Galloway 33). To start analyzing an interface as a process or translation, Galloway uses the concept of "intraface," developed by Gerard Genette in *Thresholds*, to describe the "interface internal to the interface" (Galloway 8). Galloway analyzes a still from the videogame *World of Warcraft* using the concept of intrafaces to identify the tension between two main spaces he sees in the still: a diegetic space consisting of a volumetric representation of a castle interior; and a non-diegetic space composed of progress bars, letters, and icons. Galloway observes that the non-diegetic space becomes as or more important than the diegetic space in terms of the gauges and progress bars that the user focuses on to play the game (42). In this example, the interface does not function as "a 'window' interface between this side of the screen and that side . . . but an intraface between the heads-up display, the text and icons in the foreground, and the 3-D, volumetric, diegetic space of the game itself" (42). While insightful about the dominance of the non-diegetic in this still to perform a social analysis of *World of Warcraft*, Galloway's analysis of intrafaces is also useful for a rhetorical visual analysis of conventions because he identifies the main elements in an interface-image not only for their appearance or connection to a convention, but also for how the elements interact with each other in relation to their functions for viewers.

Analyzing interfaces for how visual elements contribute to conventions or signal functions that may coordinate with or disrupt each other helps to reveal how the interactions of interface users with the other elements of the interface (such as data and the nanoscale, in data images such as STM images) are organized. Collin Gifford Brooke's observation that video gamers' complex interactions with interfaces combine and change as the game progresses suggests a closer look at interfaces in other contexts (138). Brooke's observation also suggests a closer look at interfaces beyond the common understanding of interface in purely semantic terms, and beyond the ubiquitous metaphor of interface as window. Instead, the interface is "a constitutive boundary space, not just a place of mechanistic negotiation and exchange among elements" (Drucker, "Reading Interface" 216), a space where communication is not only about signification or meaning-making, but also involves asignificatory elements in the sense of Jacques Derrida's reintroduction of the sense of communication operating in the phrase communicating doors ("Signature, Event, Context" 309).

What is particularly valuable for rhetoric about analysis of the interactions of the visual elements of an image-interface is the way the focus on elements as *contributors* to an interactive process allows for more than significatory elements to become visible in rhetorical processes. STM users and image viewers contribute to the construction of the STM image-interface: As we engage in haptic vision practices, and manipulate the components of the STM that enable response, we engage in the process of interface construction and become enmeshed, embodied, and persuaded by the process as we learn about the nanoscale. Reading STM images in terms of their intrafaces and our viewing practices thus make visible what Mark Hansen writes is one of the effects of technologies:

> By impacting our embodied lives in ways that remain invisible at the level of our culturally inscribed expectations, technologies effectively force us to experience changes in our material environment that are no longer thematizable in representationalist terms, that can only be lived through at the level of our embodiment. (*Embodying Technesis* 60)

Consideration of how form structures our interaction with images supports the assertion that STM images communicate more than significatory messages. As opposed to textual reading practices, viewers gather an impression of the images before they begin to understand the meaning of the images: At first, images function on asignificatory levels before functioning on significatory ones. Asignificatory first impressions persist, even if readers later understand the significatory messages of the images to contradict the immediate reading. For example, while STM image viewers who understand atoms in terms of quantum mechanics can impose their own understanding of atoms on STM images, STM viewers *also* perceive and respond to conventions used in images like perspective because the convention forms such a pervasive aspect of Western culture. The images thus not only organize data, but they also begin to structure readers' engagement with information experientially.

While Galloway analyzes two "spaces" that can be identified as separate elements of the image, elements such as conventions found in STM images and other digital scientific image-interfaces may overlap or occupy the same space. As demonstrated with the technique of contour that may also be seen as a landscape, it may not always be helpful to separate out modes of reading images. One may smudge into the other in surprising or subtle ways; one may work with or against another to create particular rhetorical impacts. Reading them together as components of the intraface can form an effective path of analysis for these images that takes into account this interaction.

Reading the conventions of linear perspective and those of the image-interface as elements of an STM image intraface is one way to understand Ottino's critiques with which this chapter started. In the *Science* cover image (Figure 5), the conventions of linear perspective first establish the organization of the data, so that viewers seem to be sharing the perspective of the image creator, looking into the scene of the gold wires and nanotubes. In particular, the nanotubes seem like solid objects with shadows and reflections cast on the shiny surfaces below. The bright color, the exact repetition of objects, the depicted disparities that Ottino points out, even the perfect shine—these elements create a sense of unreality that disrupts the convention of linear perspective, suggesting a haptic connection. Reading for differences in data leads to a focus on contours, on the haptic texture of the carbon atoms in contrast to the sleek gold, revealing a place of disruption or tension between the two conventions.

The "Quantum Corral" image (Figure 4) also forms a good example of intraface tension, as linear perspective is disrupted by the lurid color and even the utterly black horizon, reminiscent of the "black computer screen" that Eigler mentions.[82] While first reading the frame of the quantum corral image, the impact (the physical impact, even) of the color and dramatic perspective calls the eyes to explore. (See Chapter 4 for more discussion of this image.) The pictorial techniques of color and perspective enhance the informatic ride—the read—over such looming peaks as the iron atoms in the color version.

In reading the intraface tensions in "Molecular Switches" (Figure 6), the combination of the bright red with the spiked peaks attracts the eye. The viewer's attraction to the red peaks is reinforced by the looming, deep blue of the foreground and the flat black of the background. The unreality of the image, and its lack of participation in linear perspective, is emphasized by the delicate two-dimensional tracery of the molecular diagrams. While the dominant impression for many viewers of STM images such as "Molecular Switches" may be that of a perspectival representation of objects due to the dominance of the convention in Western culture, the form of the interface disrupts reading in terms of perspective. Instead, the conventions of the image-interface direct viewers' attention to sections of the data through color, contour, and time, conventions that draw and keep viewers' attention on the surface, as viewers make implicit or explicit comparisons to other data following the scientific convention of comparison (Bastide 197). The use of perspective in "Molecular Switches" is an example of referencing the convention of linear perspective while, at the same time, disrupting perspective through multiplication. Again, the use of multiplication fits within larger scientific conventions and expectations; yet, in "Molecular Switches," multiplication disrupts linear perspective through intraface tensions between image-interface conventions and traditional linear perspective.

Reading the tensions between conventions of linear perspective and emerging haptic conventions in the intraface helps show the dynamics of the image composition: The focus on interactions slows down the blurring between the conventions of image and interface to show some of the dynamics, including the rhetorical dynamics that create the blur. Within the STM intraface, the convention of linear perspective can be read as first organizing viewers' understanding of the data, creating associations to scientific imaging conventions. Linear perspective also acts as a filter, creating a persuasive reading of atoms as objects for anyone not well-versed in quantum mechanics. If a viewer does not bring further understanding to the image, the filter of understanding atoms as objects may satisfy the viewer. However, if a viewer also engages in haptic viewing practices with the image-interface, then color, contour, and time take the viewer along through the main argument, creating a sort of "haptic consistency" in which the viewer perceives the atomic phenomena. In turn, the convention of linear perspective helps the data seem perceivable—authentic even—by association, as readers move through the data. The feel of authenticity enhanced by perspective helps create an immersive experience amid data-become-objects. However, conventions of linear perspective argue for an authoritative frame to the image-interface that is not provided by form of the digital image.[83]

The fact that the established linear perspectival convention and the emerging haptic convention occur within the same space—within the same elements or areas of an image—and create tensions in the interface through how the conventions summon different reading practices, may contribute to their rhetorical impact, as the conventions engage readers intellectually and experientially. Jessica Hullman and Nicholas Diakopoulos's observations in their work on the rhetoric of narrative visualizations suggest the possible rhetorical impact of using more than one convention in other visual forms, such as informational images. Hullman and Diakopoulos suggest that visualizations that combine contrasting denotations and connotations are engaging precisely because of appeals to the reader through the interplay of these contrasts (2239). The back-and-forth required to engage in both perspectival and haptic viewing practices in the STM image-interfaces may create a feeling of interest and, as mentioned above, create a sense of the data—and nanoscale—as perceivable and real. Returning to Drucker's articulation of the interface as a constitutive boundary space, what the process of reading conventions as intraface elements highlights is how the scientific image-interface constitutes and is constituted not only by data, viewers, STM apparatus, and atomic phenomena, but also by conventions—and, as part of conventions, reading practices and the rhetorical possibilities that intrafacial dynamics reveal.

Conclusion: Haptical Consistency and the Shifting Possibilities for Rhetorical Analysis

Returning to the example of misreading images that began the chapter, Ottino's reasons for worrying about the accuracy of scientific images are based on the ability of images to persuade viewers that what images depict is or could be reality, and that images create excitement. Ottino brings up two of the main ways that STM images communicate, both of which emerge in the discussion above: imaging conventions, particularly those associated with linear perspective, help persuade viewers that what the images show is indeed reality as they help readers recognize the familiar in the nanoscale (although readers' perceptions of the images do not always align with quantum mechanics conceptions of atoms). Also, the haptic cues (that should be considered a shift in conventions) arrange viewers' experiences of the data, as viewers use haptic vision practices; the tactile, experiential reading practices used by STM image viewers, combined with the authentic "feel" of the data from linear perspective conventions, generate excitement and engagement on the part of the viewer. The interface as process or constitutive boundary forms a critical part of the creation of meaning in a STM image-interface.

While the interactive practices used to create STM images and the haptic viewing practices used to understand data help to create the blurred boundary between pictures and data, still images, and interfaces, the interactive, haptic practices also blur boundaries between viewer, data, and nanoscale within the space of the interface. Along with data, viewers, vision practices, and nanoscale phenomena constitute and are constituted by interactions that compose the interface; in the interface, then, rhetorical possibilities expand as the elements that create the interface interact.

The fact that STM images show some conventional disruption suggests changes in reading habits or discourses; the emergence of such disruptions, combined with the different affordances of interfaces—including interface functions and the capacity to turn viewers into users—suggests a shift in rhetorics as well. Rhetoricians have always analyzed form and content together; in the case of digital images, analysis of form and content is particularly important. As Drucker observes, form often *is* the information (*Speclab* 17). My analysis of shifting conventions suggests some attendant shifts in rhetorics to account for the ways that experiential, procedural, and visual communication made with digital visualization technologies like the STM are persuasive.

A focus on an analysis of the interface, for example, helps account for how images argue for the tangibility of atoms. As I discuss above, perspectival conventions used in combination with emerging data image-interface conventions help structure the concepts of atoms as well as the possibilities of interacting with atoms as we would with any other object: We can move atoms around, touch them, and expect the atoms to behave predictably. Analysis of images as events through interface dynamics helps account for the claim of communicating manipulative possibilities. The fact that STM images function in both significatory and asignificatory ways to create an immersive experience shifts the rhetorical focus away from analysis of a product and towards analysis of an event, process, or procedure. STM image-interfaces create a sense of tangibility not entirely because the image-interfaces seem to represent objects; STM images also create a feel of manipulability because they function as interfaces that allow the viewer to become the user and to manipulate data (in the case of images) and atoms (in the case of images that are part of an STM apparatus, as Chapter 1 discusses).

Expansion of the rhetorical analyses of procedures, of processes as rhetorical "texts," or perhaps as events (as Kevin Michael DeLuca argues for analysis of images) is one way to account for some of the work of interface-images. As discussed in Chapter 1, Ian Bogost's formulation of procedural rhetorics in videogames provides a useful way to understand how situations in which users and viewers engage in other structured experiences, such as those in STM image-interfaces, can be rhetorically analyzed. Bogost's focus on how the responses of users are shaped by and help shape the rhetorical messages of games provides a useful map for identifying other procedural rhetorics, such as those of scientific visualizations. Further assessment of rhetorical patterns and procedures within other image interfaces could build on the analysis of STM image-interfaces presented in this chapter.

The study of the dynamics of visual conventions in this chapter—and of dynamics that are not entirely signficatory—suggests that assessments of the rhetorics of the scientific image-interface need to include significatory representation, but not be limited to significatory representation alone. Part of the task of analyzing interfaces, as Drucker articulates, is that we need to "understand how they organize our relation to complex systems (rather than how they represent them)" ("Reading Interface" 213). In fact, a focus only on representation in the case of interfaces can be misleading. Drucker observes that the immersive quality of interfaces, such as those developed by virtual reality experimenters and GUI developers, does not require full replication of the sensory world to function. Further, the degree to which an interface might seem to correspond with reality is not a good indicator for how well the interface works. Drucker cites Microsoft's failed interface, "Bob," as an example ("Reading Interface" 214). Therefore, what is important for understanding the rhetorical nature of STM images is not strictly what the images contain, but how they contain what they do: how STM images position the viewer to perceive and respond.

My account of the disruption of conventions of linear perspective raises another question about the study of scientific visualization in regard to the concept of "optical consistency" that, as Ivins and others argue, forms a key component of scientific communication. For Ivins, the conventions of linear perspective allow for the creation of optical consistency, enabling anyone who sees a depiction of an object using such conventions to be able to recognize that object. It could also be argued that conventions associated with haptic viewing practices may form their own consistency—a haptic consistency, perhaps, that provides a systematic way to convey qualities of manipulation and interaction that STM images express along with the data. While the convention of linear perspective has not been abandoned, and, instead still functions in STM images, the fact that linear perspective does become disrupted points to the significance of those disruptions. As conventions change over time, as with language, will haptic viewing practices generate a sense of consistency that allows for images and other visualizations to function with the same back-and-forth capacity that makes optical consistency such a powerful tool?

The focus on convention also brings up a key danger that Ottino's commentary suggests: that of exceeding the bounds of convention so much that images become less credible to their audience. Elkins recounts the danger of disrupting conventions too much, or of getting rid of conventions altogether in a scientific context:

> When scientific images begin to work as art does—that is, by giving up secure meaning in favor of a halo of pictorial possibilities—then their place in science becomes problematic. Even though their popularity as teaching tools, and even as guides for research, might increase (such is the power of a well-constructed image), their utility might become unpredictable. (*Domain* 38)

While STM images do not seem to function completely as art, the possibility that STM images might is occasioned by new visualization technologies. The connection of an image to its audience in light of pictorial conventions is one reason why the study of scientific images in the context of their digital and interfacial characteristics is important: The use of new media, such as digital media, creates possible problems for science communication in that some of what might seem particularly strong rhetorical choices to make in images–such as the use of linear perspective, for example—may be disruptive to other conventions at play in the image, depending on the reading practices of viewers. The analysis in this chapter is a start at understanding the ways that visual conventions interact, especially as components of intrafaces, to better understand the rhetorical work of STM images and of the interface as a form, and of the interface as a scientific visual inscription. Chapter 4 continues this study of visual rhetoric and discursive change from a different angle, as I analyze common scientific tropes that appear in STM images and nanotechnology to further explore how haptic reading practices manifest.

4 Visual Intelligence: Reading the Rhetorical Work of STM Images in Tropes

According to one dominant narrative,[84] the conception of nanotechnology as a field occurred in "There's Plenty of Room at the Bottom: An Invitation to Enter a New Field of Physics," Richard Feynman's December 29, 1959 speech given at the annual meeting of the American Physical Society. In this speech, Feynman proposes a future field based on "manipulating and controlling things on a small scale," arguing that no theoretical reason contradicts the possibility of making storage devices and other tiny machines out of atoms. At the time, as audience member Paul Schlicta recalls, most of the audience thought that Feynman "was trying to be funny" by presenting a series of zany propositions (qtd. in Appenzeller 1300). Feynman's talk rapidly gained wider attention. For example, the speech was reported in magazines such as *Popular Science*, *Life*, *Science News*, and *Saturday Review*; attracted competitors for the two prizes he offered at the end of the speech; and was subsequently published in *Engineering and Science*'s February 1960 issue, in a book entitled *Miniaturization*, and was excerpted in *Science*'s nanotechnology issue (Regis 72–73, 75). It now also appears online.[85] Feynman's speech gained attention as a founding document of nanotechnology after K. Eric Drexler cited it in a 1986 promotion of nanotechnology for a popular audience, *Engines of Creation* (40–41); since then, "There's Plenty of Room at the Bottom" has assumed a nearly uncontested founding role in origin narratives of nanotechnology.[86]

Feynman's speech has remained a touchstone for nanotechnology in part because Feynman communicates his ideas to his audience efficiently and powerfully through the use of a few figures of speech common to scientific discourse, including the tropes of comparing objects to tiny writing associated with "the book of nature," and of what microscopes show as subvisible new worlds. These tropes articulate the possibilities and boundaries of the "room at the bottom" within Feynman's speech, and the tropes also link Feynman's ideas to the established scientific fields in which the tropes occur. Subsequent visual and textual citations of Feynman's speech have also helped these tropes become common in nanotechnology discourses.

The use of tropes that refer to writing in the book of nature and subvisible new worlds helps establish rhetorical conventions that can occur in both image and text. Tropes thus help to shape discourses of nanotechnology as well as knowledge about the nanoscale. This chapter explores how the image-making and vision practices solicited by the STM affect reading practices through analysis of how the tropes of tiny writing associated with "the book of nature" and of subvisible worlds appear in STM images. The appearance of these tropes in written discourse and images supports arguments about the definition of nanotechnology by communicating throught the use of concepts that are more familiar than quantum mechanics to the images' viewers: letters and landscapes. Additionally, the repetition of the tropes argues in less significatory ways by reinforcing viewing practices associated with haptic vision, and so help constitute the content and parameters of nanotechnology discourses.

To follow how STM images help shape concepts of the nanoscale, this chapter analyzes STM images in which the use of tropes of writing in the book of nature and subvisible new worlds summon associations with conventional uses of the tropes in scientific discourse; yet, the tropes also create other associations that are influenced by the interactive production and vision practices necessary to make and read STM images. Following these tropes can show how and where the tropes contribute to arguments, affect scientific literacy practices, and affect understandings of the nanoscale and nanotechnology. First, however, a few words about metaphors can further explain why following tropes can illuminate aspects of the development of nanotechnology.

Shaping Communication, Shaping Knowledge: Tropes, Images, and Nanotechnology

Scholars have long studied the influence of tropes, particularly metaphors, on shaping communication and knowledge in science, often building on the work of Max Black, who, with *Models and Metaphors*, re-energized consideration of how metaphors function to communicate new ideas. Metaphors do so by building associations between the subject being described (the tenor) and the subject that is associated with it (the vehicle). Black argues that not only is using one subject to illuminate another important mental exercise, but that this exercise demands "simultaneous awareness of both subjects but not reducible to any comparison between the two" (46). Additionally, drawing from I. A. Richards's *The Philosophy of Rhetoric*, Black argues for an interactive view of metaphors in which metaphors are transformative of both subjects. Following Black, scholars such as Theodore Brown, N. Katharine Hayles (*How We Became Posthuman*), Evelyn Fox Keller, George Lakoff, and Mark Johnson, for example, argue that figurative language such as metaphor plays a critical role in creating mental models, envisioning ourselves and objects, and shaping fields of knowledge—including scientific fields. Indeed, Mary Hesse suggests that when applied to science, Black's interaction theory of metaphor helps create new knowledge (123).[87]

The influence of metaphors on understanding exists partly because of the ability of metaphors to link complex, sometimes non-intuitive, or new ideas in science. As Arthur I. Miller observes, the concepts or domains that metaphors link are often quite complex in their own right, and "are partially represented in non-propositional modes of thought, such as mental imagery" (147). The ability of metaphors to convey non-propositional concepts adds to the linguistic capacities of language to express new or complex ideas. Indeed, Brown argues that metaphors help to shape scientific thought not only in theory, but also in experimental science. Brown describes the role of metaphor in experiment: "There is an interaction between the metaphorical frame of thought and the literal observation system in which experiments are performed. An apt metaphor suggests directions for experiment. The results of experiment in turn are interpreted in terms of an elaborated, improved metaphor or even a new one" (26). Therefore, exploring extant metaphors or other tropes can help determine how a particular field or concept is framed within scientific discourse.

Metaphors shape our understanding by focusing attention towards certain aspects of the object or idea under scrutiny, and away from others. As Black explains, the reason for directing attention is not only to emphasize or suppress qualities of the original object, but to also organize them so that the metaphors behave, as Kenneth Burke describes, as "terministic screens" (44-45). This organizing dynamic can create negative consequences, for metaphors also constrain thought. James Edie, for example, argues that once people "have unwittingly chosen their metaphor, they tend to think within the cultural-linguistic bounds that they have unwittingly set up for themselves" (171). The tendency of people to unreflectively use metaphors as guides for thought explains some of the power of metaphors in shaping thought and knowledge.

Another reason to look at metaphors closely is that metaphors can shift depending on cultural contexts. Because metaphors rely on a familiar or given subject—the vehicle—with which to describe a new or complex term or phenomenon, as Black explains, metaphors only work as long as the associations they evoke are common and readily brought to mind by readers (40). Therefore, tropes may shift readily. For example, Brown asserts that metaphors associated with global warming changed as they moved beyond scientific discourse to wider contexts (179). Following this view, tropes may also shift within new scientific contexts, given cultural or scientific changes. Therefore, following tropes with attention to how tropes change provides insight into how figures may not only impact discourse in a given field, but also how the fields in which tropes are used may be developing.

While most research on metaphors focuses on linguistic metaphors, a case can be made for the study of metaphors and other tropes that exist in visual images, especially in science.[88] Following the line of thought that metaphors and other tropes are conceptual and help to shape knowledge suggests that images, such as images of nanotechnology, may also contain metaphorical material.[89] Some researchers of visual images in science have indeed found rhetorical structures in scientific images. For example, Jeanne Fahnestock finds that scientists often use visual figures in their arguments, and even use the same kinds of rhetoric in both text and visual figures at times (xi). Luc Pauwels discusses metaphors as "conceptual structures" in his framework for understanding scientific images (3). Brigitte Nerlich argues that visual images themselves function like metaphors in that the images do not only bridge between the new and the unfamiliar, but also "invite expectations and reactions" (289). Viewing figures as conceptual, not just linguistic, structures opens up the way to more closely examine how images may function in—and as—scientific communication.

Analysis of metaphors and other tropes can be particularly useful for understanding the development of emerging fields like nanotechnology, especially (as discussed in the introduction) due to the stage of emerging technology development that emphasizes communication and negotiation processes (Fiedeler 248). Indeed, the ability of metaphors not only to explain complex ideas, but also to help develop understandings of new ideas, gives metaphors great explanatory power and potential in nanotechnology, where all phenomena are not verifiable by the unaided eye.[90] Additionally, as the introduction discussed, the definition of nanotechnology is a contested, complex, and evolving issue.[91] Brown's explanation of the role of metaphor in the experimental process may also be an apt application to how metaphors help develop a field of knowledge (26). The fact that nanotechnology is still emerging suggests that metaphors and other tropes may play a larger role in shaping the concept of nanotechnology than a field that is more established.

Metaphors and other tropes have great explanatory power and potential in nanotechnology in particular, especially given that quantum physics is not easily explainable using language. However, a few factors make using tropes to describe science a complex situation, including (as mentioned above) the complexity of the concepts metaphors link to and the risk of constraining ideas about novel phenomena. In addition, the ambiguity of tropes makes them complex: While the fact that the meanings of metaphors and other tropes are not narrowly circumscribed may be an important way that tropes help images, such as STM images, form instances of "wordless creole," part of a trading language between different disciplines or groups (Galison, *Image and Logic* 52), or instances of productive ambiguity between or even within disciplines (Keller, *Making Sense of Life* 113-98), such ambiguity can be misleading. Audiences of tropes may develop different perspectives on the same point, similar to how image viewers may develop different ideas through misunderstanding the visual conventions, as discussed in Chapter 3. For example, differing readings of tropes create alternate understandings of the nanoscale and of nanotechnology. Differing readings become particularly significant considering that the definition of nanotechnology, to which the tropes discussed in this chapter can contribute, is still forming. How tropes such as those that refer to writing in the book of nature or subvisible new worlds are read, and what the tropes suggest about the nanoscale, can then affect arguments about the possibilities and future directions of nanotechnology.

Reading Nanotechnology: Atomic Writing/Writing Atoms

Images of the nanoscale that have influenced the definition of nanotechnology the most include D. M. Eigler and E. K. Schweizer's "IBM" images (Figure 1 and Figure 2). Eigler and Schweizer's "IBM" images (and the text descriptions of the images) circulated widely throughout scientific and non-scientific venues, including newspapers, television, popular magazines, and scientific journals.[92] Many citations of Eigler and Schweizer's images appear in the context of documenting the emerging origin narrative of the development of nanotechnology, in which Eigler and Schweizer's manipulation fulfills a prediction outlined in Feynman's speech. Repetitions of the "IBM" image do more than spread the news about Eigler and Schweizer's experimental results; through proliferation and use, citations of the "IBM" image also help create what is considered nanotechnological fact, according to Bruno Latour's description of how scientific statements become fact. Latour explains that citations help convert the claims of scientific statements into facts, because each citation does not call into question the original experiment; instead, the act of citing in the service of another argument causes the claim to be treated as a fact on which the citing authors build their own argument (*Science in Action* 38–42). Repetitions of the image, then, as they appear in other arguments as well as in their original role within Eigler and Schweizer's publication, help shape scientific and non-scientific understandings of nanotechnology and its possibilities. Therefore, what the "IBM" images make visible and possible affects the formation and discourses of nanotechnology.

Eigler and Schweizer's images go beyond mere nano-advertising: They also function as proof that such manipulation has occurred, following the use of uniquely human patterns, letters, in the image. Writing in the "IBM" images also functions as a metaphor for manipulating atoms, and links to other figurative associations with writing—such as the common, scientific trope of the book of nature—that circulate through nanotechnology, including in Feynman's speech. Reference to the conventional scientific trope of writing contributes to the force of communication in the "IBM" image, and also to Eigler and Schweizer's new abilities through the presence of the trope within existing scientific rhetorical patterns, therefore increasing the credibility and clarity of the "IBM" image through the use of familiar forms. At first glance, the use of the trope might appear to fit neatly within common scientific figurations of writing in the book of nature; indeed, the trope can be read as such. However, how Eigler and Schweizer present writing in the "IBM" images suggests a reading of the letters that the images portray that deviates from many instantiations of the trope in science, including its appearance in Feynman's speech.[93] A comparison of Feynman's use of the trope of the book of nature and its appearance in Eigler and Schweizer's images reveals a shift from configuring writing as a product to writing as a process. I argue that part of this shift is due to the deeply interactive processes associated with using the STM and producing STM images, as well as the viewing practices STM images solicit. The shift of this trope from writing as product to writing as process is significant because the writing in the "IBM" images makes atom manipulation—and atoms as objects of scientific knowledge—visible in ways that complicate the current narrative of the development of nanotechnology as well as arguments about the definition of nanotechnology and its future possibilities. The divergent readings of the trope in terms of writing as product or process also indicate a possible point of disjunction in the formation of the field itself.

Making Matter Signify: From the Book of Nature to Writing Atoms

In "There's Plenty of Room," Feynman first mentions writing to clarify the scale he envisions when he says that there is room at the bottom. Feynman explains that "there is a device on the market . . . by which you can write the Lord's Prayer on the head of a pin. But that's nothing; that's the most primitive, halting step in the direction I intend to discuss" (22). He asks, why can't someone "write the entire 24 volumes of the Encyclopedia Britannica on the head of a pin?" and suggests writing the holdings of the Library of Congress, the British Museum Library, and the National Library of France on one pinhead (22). Feynman not only refers to micrographia—the practice of tiny writing[94]—but also, in his expansion to the encyclopedia and then to not one but three libraries, his own evocation of writing approaches the writing more closely associated with the trope of the book of nature.

"The book of nature" configures the world as composed of signifiers waiting to be understood by humans who can read this natural language. The figuration of nature as a book appears in ancient Greek texts, in which the trope of alphabetic writing describes concepts such as atomism and the four elements (Hallyn 53). The trope emerged in medieval European sermons and middle-eastern texts as an analogy between the Bible and the natural world as God's writing (Eisenstein 456). In medieval Europe, while the book of nature was construed as a series of resemblances that were understandable if read allegorically, in the seventeenth century another "book" emerged in addition to the first. This second book of nature entered scientific discourse as a favorite rhetorical figure among early-modern scientists such as Francis Bacon and Galileo Galilei: It suggested that measurable facts and laws of nature were presented to those who could read them empirically (Robinson 89). In what is perhaps the most famous description of the book of nature, Galileo, for example, contended that the "grand book the universe which stands continually open to our gaze . . . is written in the language of mathematics . . . its characters are triangles, circles, and other geometric figures" (qtd. in Eisenstein 458). The suggestion that humans can not only "read" the wonders of the natural world, but can read them in the language of mathematics became a guiding and justificatory theme in scientific discourse, as it associates reasoning through mathematics with the facts of the natural world. In Feynman's speech, the trope links the writing he envisions with objects such as atoms that already reside in the book of nature. In connecting natural objects with writing, Feynman's use of the trope prioritizes "writing" as a noun—a physical collection of information for people to read if they know how.

Feynman's reference to writing, with its links to the book of nature, relies on a specific concept of information. After invoking the idea of tiny text, Feynman soon shifts from letters to what he calls the "information content" of letters (24). He envisions recording "information content in a code of dots and dashes or something like that, to represent the various letters" by using "the interior of the metal as well" as the atoms themselves (24). The writing trope allows Feynman to shift from recording information (i.e., collecting the contents of the book of nature) to writing information, rearranging the "natural" material to the human-made by moving atoms into a pattern that records data. Feynman's version of writing, with its associations to the book of nature, enables the shift between recording to writing information to occur seamlessly, as he moves from the representational—letters, similar to Galileo's geometric figures and mathematical symbols—to the material—atoms. Yet, the shift from recording to writing is only possible through the definition of information that Feynman implies in his phrase "information content": By emphasizing the ability to separate content from other aspects of information (such as form), Feynman's information aligns with Claude Shannon's definition of information as a substance-free pattern, as articulated in Shannon and Warren Weaver's, *Mathematical Theory of Communication*. Information as pattern emphasizes significatory qualities over more material ones, and lets Feynman envision atoms as stabilized entities so that the atoms can be "read" and "written"—and so make meaning for the reader—while glossing over consideration of possible material effects of atomic rearrangements (such as on neighboring atoms, or on the instruments manipulating the atoms). Even as Feynman conceptualizes atoms as elements of a material-free pattern, he also describes a physical place as a storage place for "information content." Feynman's description of storage has spatial dimensions, and is yet physically alterable—a point that Feynman elides, perhaps because of the extremely small size of atoms.

The writing trope resurfaces later in Feynman's speech when he explicitly connects writing to the natural world:

> [T]he biological example of writing information on a small scale has inspired me to think of something that should be possible. Biology is not simply writing information; it is doing something about it. A biological system can be exceedingly small. Many of the cells are very tiny, but they are very active; they manufacture various substances; they walk around; they wiggle; and they do all kinds of marvelous things—all on a very small scale. Also, they store information. Consider the possibility that we too can make a thing very small which does what we want—that we can manufacture an object that maneuvers at that level. (25)

In this instance, the trope of writing allows Feynman to draw an analogy between the writing within the book of nature and human writing, echoing Galileo's reference. Feynman expands from writing to the effects of writing, "doing something about it." Through the analogy of the book of nature's storage of writing, Feynman suggests that because it is not a biologist, but is biology that originally manipulated information, the scientists he foresees manipulating at the nanoscale would use the same process already involved with the biological writing in the book of nature. Here again, the concept of information inherent in Feynman's analogy is information as a material-free pattern, able to be stored like a tiny version of the Library of Congress, no matter whether the information is written in text, images, or nucleotides. Feynman's association of manipulation with nature erases possibilities for effects on the writer, as this version of writing assumes a stable, unchanging, and distant writer having written in the book of nature, manipulating without contact, without being touched in return.

Therefore, the use of writing in Feynman's speech draws from the book of nature figuration to make matter signify while, at the same time, downplaying the very "mattering" of matter. Because of the conception of writing the trope employs, Feynman's ideas of what the "room at the bottom" looks like, and how this room could be used, rely on dematerialized information flow and storage. Feynman's use of the trope of writing in the book of nature helps frame manipulation in terms of conventional scientific conceptions of human relationships with nature as a book written in objects and mathematics. Feynman thus articulates—and anticipates—atomic manipulation as a scientific practice in which scientists maneuver individual atoms to store material-free information without affecting the scientists who perform the manipulation. Following the logic that the conventional trope of writing in the book of nature implies, emphasis on material-free information would also allow the writing in Eigler and Schweizer's images to appear as simply revelation, not production; however, this reading omits important communication about the nanoscale that Eigler and Schwiezer's images convey.

From Information to Act: Writing IBM

Feynman's call for the invention of new microscopes in order to image the "room" he envisions (24) links to another formative event in the common narrative of the development of nanotechnology: Gerd Binnig and Heinrich Rohrer's invention of the STM. According to the common story, after the invention of the STM, Eigler and Schweizer heeded Feynman's call—they used the STM to manipulate atoms.[95] Feynman's speech and Eigler and Schweizer's "IBM" images are linked not only by appearing in an origin narrative of nanotechnology, but also by using writing as a structuring trope. At first glance, the trope of writing seems to function in the "IBM" images as it does in Feynman's speech. Indeed, one could read Eigler and Schweizer's choice of writing to demonstrate the manipulability of atoms as a reference to the book of nature, as following biology's lead in what Feynman characterized as "doing something about it" (25). (Literally, of course, Eigler and Schweizer's images appeared in the journal Nature, if not the book of nature.) In addition, reading the use of writing in the "IBM" images in relation to the book of nature figuration emphasizes the textual—not the imaged—form of Eigler and Schweizer's accomplishment. Reading "writing" in relation to the book of nature trope blurs the line between text and image form, enabling the images to be cited as text. Additionally, in written descriptions, such a reading smudges the differences between image and text, because such writing necessarily includes the main entity of the image in visual form on the page: IBM.

The appearance of the trope of writing in Eigler and Schweizer's images strays from Feynman's use in a few important ways. The reading practices associated with emerging haptic conventions that Chapter 3 discusses, as well as the interactive imaging and viewing processes that Chapter 1 and Chapter 2 describe, create a different view of what the "IBM" images communicate, one that de-emphasizes the view of objects as fixed examples of the writing of nature, and therefore revises the link to the book of nature figuration. Reading, imaging, and viewing practices comprise parts of the network that form "the facts" Eigler and Schweizer's images communicate; and so, when the "IBM" images are read following viewing and reading practices associated with the STM, and in light of the form of the image-interface, the writing appearing in the "IBM" images becomes linked to a highly participatory process—not a collection of information such as the writing found in the book of nature.

The Work of Images

That writing in the "IBM" images does not occur within a speech or written document (like Feynman's use of the trope), but within an image, changes how viewers—even those not used to the STM—respond to the reference to writing. For example, because of the use of a shadowing effect under each atom in the "IBM" images that creates associations with the perspectival convention, viewers first get an impression of raised bumps like an atomic Braille that viewers may feel on their fingertips as they run their eyes over the surfaces of the image. Viewers then connect the dots together, like a constellation of stars, to read the letters I, B, and M together. Only then do viewers, as readers, register the atomic writing as the name of the company, a sort of atomic graffiti: "IBM was here."

Because images create pre-significatory, affective responses, as Steven Shaviro argues (26), the images have already communicated—and viewers have already started to respond—before processing the meaning the images convey. Affective responses constitute one major difference in how images and texts communicate: Communication in images functions as a force, transmitted through affective contact between the image and the viewer, conducted more by a "tremor, a shock, a displacement of force," as Jacques Derrida describes communication, and not only by signification ("Signature, Event, Context" 309). Like other visual forms, the "IBM" images affect viewers even before viewers sense the meaning the images convey.

In showing spaces between the atoms, Eigler and Schweizer's images also extend what is usually the instant separating seeing what text is made *of* and understanding it *as* text. As viewers of the "IBM" images begin the process of understanding what the images signify, viewers' reading process becomes more participatory. Viewers must first mentally assemble letters from the collection of atoms through the pattern the atoms form, as dramatized in the series of images accompanying the *Nature* article. Viewers do so by running their eyes closely over the texture of the bumps, participating in haptic vision practices. When viewers finally see the arrangement of the atoms as letters, the "IBM" images mimic a page or word-processing program so that viewers see the letters as text. While viewers respond viscerally to the images, and form the letters that they will then be able to read, viewers become experientially involved in the process of reading, a process that also creates an impression of atoms as tangible entities, as viewers move from atom to atom to assemble the letters.

The combination of presenting Eigler and Schweizer's accomplishment in a series of images, creating a more participatory reading process for viewers of the "IBM" images, emphasizes the interactions necessary to read the "IBM" images. The emphasis on interaction counters the concept of reading inherent in the conventional use of the book of nature figuration. Readers of the "IBM" images do not merely try to understand the signification of natural marks (here, atoms); instead, readers also construct the writing as they read the letters in the images. The participatory process of reading changes the connotations of writing in the "IBM" images, and, in so doing, allows the image to create a different relationship of the reader to writing than the manifestation of the trope in Feynman's speech.

Writing as a Process: STM Imaging

The production processes of STM images contribute to the participatory reading process required to read the "IBM" images because the

STM user "sees" through the computer screen, and interacts with the nanoscale sample through the onscreen image. The image becomes not only a display, but also an interface with the sample, changing the reading practices that STM users engage in when they read STM images, as previous chapters have outlined. As Chapter 1 discusses, in September 1989, Eigler and Schweizer noticed a few streaks in an image the STM produced during an experiment with xenon. While investigating the cause, Eigler and Schweizer realized that they could create the streaks while maneuvering the tip of the microscope (Eigler, "From the Bottom Up" 431). The two scientists experimented with moving a xenon atom, whose progress they followed on the screen of the STM by flipping back and forth between imaging and movement modes. After some practice in manipulating the xenon atom on the nickel surface through interacting with the imaging and tip-positioning capacities of the STM, Eigler and Schweizer generated another image of xenon atoms—this time, the atoms were lined up in a row. They then produced the "IBM" images (Eigler, "From the Bottom Up" 431). In the process of producing the "IBM" images, Eigler and Schweizer learned how to respond to the atoms and the microscope in particular ways through interacting with the images—images the scientists then turned into writing. For the STM users-become-writers, writing is similar to the process of interacting with atoms, as a pattern emerges from the movements of the scientists' hands, shifting into letters and into meaning. Writing becomes a strategy for learning how to interact with atoms.

The participatory nature of the STM imaging process is reinforced through the continued processing of the images after the data is collected and displayed as a matrix. While many researchers display the image the STM produced "as is" in journal articles to faithfully present the data, to create images or figures that highlight certain aspects of the data, or to create color images for journal covers or websites, researchers interact with data points using the computer's ability to transform data into pixel brightness on the screen of the STM, using programs like *MATLAB* or *Photoshop* to emphasize certain patterns of collected data. Therefore, the writing used in the creation of some STM images links to acts that not only create new text out of random atoms, but also create an image of text. The creation of text from both atoms and pixels in the "IBM" images emphasizes the process of manipulating the atomic writing that was demonstrated in Eigler and Schweizer's breakthrough.

Emphasizing the intertwined acts of writing and atom manipulation shifts the associations of the trope of writing from writing found in the book of nature (where writing already exists and is fully formed) to writing as a repeatable, participatory act—where writing names the act that incorporates atoms as well as readers who are also writers and writers who are also readers. The association of writing with actions implies that what is to be read does not exist before the interactive process; but, as the "IBM" image viewer constructs letters from the atoms, what is to be read emerges *from* the interaction. In direct contrast to the writing trope grounded in the book of nature figuration, writing in the "IBM" images connotes more of an experiment than a collection or product. Writing in the "IBM" image encourages scientists to engage in the familiar process of writing to learn how to manipulate atoms with the STM.

Informatic Atoms, Haptic Information

While focusing on the textual qualities of the "IBM" images alone produces some sense of how the "IBM" images communicate, the deeply participatory processes that viewers and microscope users engage in also contributes to a different connotation of writing in Eigler and Schweizer's images. Indeed, the writing trope expressed in the "IBM" images contradicts the use of writing associated with the book of nature figuration. For STM users and viewers of the "IBM" images, writing becomes a haptic strategy linked to atom manipulation: Writing becomes a repeatable, deeply participatory and synesthetic act that bodily incorporates image viewers and STM users within the writing process.

The shift from writing associated with the book of nature, such as that used in Feynman's speech, to writing as a process also affects the concept of information associated with writing. Here, information becomes mutable for image viewers as readers and writers, and for STM users who both read and write atomic letters. For image viewers, the "IBM" images affect them throughout their viewing, before making meaning of the dots scattered across the screen, while forming letters out of the dots, and then in reading the letters. As STM users engage in the process of interacting with the STM sample, the displayed information materially affects users as they respond to the image of writing and the arrangement of pixels on the screen. The responses of STM users, then, alter the information on the STM's computer screen. Therefore, what becomes information for both image viewers and STM users is more tangible than Shannon's material-free patterns, as viewers move from atom to atom to create the information, and as STM users manipulate atoms and experience the effects of movements by seeing changes in the images. Information is embodied and inseparable from its form; information is also inseparable from the responses of viewers and users. Information emerges from the interactions among atoms, the STM apparatus, the form of the image, the image viewers and the microscope users in a process that suggests Susan Oyama's description of information that does not exist *a priori*, but instead co-develops with the form in which the information is communicated (2). As form and information co-develop and interact with each other in the process of formation, both the information and its presentation transform due to these production processes, contributing to the altered trope of writing.

The shift in the writing trope also calls attention to the existence of atoms as material, tangible, and manipulable objects. Unlike its use in Feynman's speech, the writing trope in the "IBM" images enables viewers to more easily connect atomic writing to other objects that may be formed from atoms (due to the emphasis on atoms as objects as well as emphasis on what objects the atoms form; here, letters). Eigler and Schweizer's imaging of xenon atoms as fixed, individual elements of the letters "IBM" presents atoms as material, manipulable entities. Reading the "IBM" images and their citations with attention to the image form and the reading practices of the viewers of the images reveals how the "IBM" images contribute to arguments about the material properties of atoms, affecting the formation of nanotechnology.

Reading Nanotechnology

The altered trope of writing exhibited in the "IBM" images also echoes in discourses of nanotechnology through the repetition and continued citation of the images. The trope thus helps to constitute nanotechnology. Both Feynman's speech and Eigler and Schweizer's images function as touchstones within arguments about nanotechnology, and so serve to not only justify the value of the knowledge the speech and images convey, but to also justify the value of images as data conveyances. Through the expression of writing as a verb, as an act involving the user and the changing relations between user and atoms in the process of writing "IBM," writing adds to one implication of nanotechnology: that the world is not a fixed text, but is manipulable to its very atoms. Also in the trope's emphasis on writing as a process, not a product, the "IBM" images train viewers to anticipate a process-based participation with the world, atoms, and images.

The participatory—and anticipatory—nature of atom manipulation expressed in Eigler and Schweizer's images also emphasizes the claim of the existence of atoms and their materiality as tangible, isolable, and manipulable objects, in opposition to Feynman's elision of the material presence of atoms in his concept of information storage. The claim of the materiality of atoms is especially communicated in the visual articulation in the image of atoms as bounded spheres (like marbles), as well as the arrangement of the atoms into letters. The articulation of atoms as tangible obscures other attributes of atoms. For example, what the "IBM" images do not make visible is that, according to quantum mechanics, atoms do not have such demarcated boundaries, but exchange electrons with neighboring atoms. What is also hidden is that Eigler and Schweizer's atomic manipulation was carried out at four degrees Kelvin to keep the xenon atoms in place long enough for Eigler and Schweizer to manipulate them. The appearance of dots that form letters creates a simplified version of atoms that allows viewers to more easily connect the atomic writing expressed in the "IBM" images to other objects formed from atoms, and so contributes to reading other images that portray atoms in similar ways.

One particularly vivid example of the implications of reading atoms as tangible entities based on the use of the trope of writing in the "IBM" images appears in images of molecular gears that Drexler posits and simulates on a computer ("Differential Gear"). There, atoms resemble differently colored marbles neatly stacked on top of each other. Eliding the participatory dynamics of atom manipulation that the "IBM" images communicate through writing obscures the controversy around whether, in fact, Drexler's gears are physically possible. These are debates that have occurred since the publication of *Engines of Creation*, and in which a few scientists, most notably Nobel laureate Richard Smalley, participated.[96] However, following the viewing and imaging practices that the "IBM" images solicit—and that the altered trope of writing here communicates—reveals the deeply interactive, participatory processes of nano-technological "writing" that incorporate viewers and microscope users, as well as atoms. These participatory processes might, for example, cause viewers of Drexler's simulations to view the images in terms of the process of how the atoms fit together, or how the simulations may move, as the viewer slides her or his glance from one atom to the next, as opposed to what object the atoms seem to form. A participatory viewing practice would thus lead viewers to ask questions about how the atoms interact, as opposed to only for what the larger object is used.

In a discussion of the competing connotations of the trope of the book of nature in the seventeenth century, Ken Robinson explains that what changed from one idea of this trope to the other is that "the book which they [readers] see is quite different. It is determined by their reading tools" (89). One way to understand the shifted trope that the "IBM" images create is to see the images as evidence for a shift in reading practices, as the tools—the ability to move atoms—create different relations between atoms, viewers, microscope users, and the microscope apparatus. The shift, then, creates a different reading practice, and leads to a different trope. Feynman may have anticipated manipulating atoms within the familiar rhetorical frame of reading the book of nature ascribed to by scientists such as Galileo Galilei; however, Eigler and Schweizer ended up departing altogether from the conventional book of nature, and instead created different reading and writing practices. What becomes visible through the practices associated with STM use is something that Feynman's frame does not show: a relation to atoms and matter that causes both STM users and viewers to not read a book, but through deeply participatory processes of atom manipulation and image production, to *co-produce* the writing STM users and viewers can then read.

The use of writing in the "IBM" images, then, does not only alter the writing trope within these images, but the altered use of writing also points to a mid-stream change in nanotechnology, occurring at the point where what is being discussed is not speculative (as Feynman's articulations were), but is experiential. Reading the "IBM" images complicates the usual narrative of nanotechnology, for reading with attention to the trope of writing points out a discontinuity between oversimplified cause-and-effect plotlines involving Feynman's speech and Eigler and Schweizer's "IBM" images. However, reading the trope of writing also suggests that a very different story is being written, one missed when considering the trope in these images as a reference to the more conventional version of writing in the book of nature. The alternative narrative of nanotechnology takes into account the materiality of the nanoscale and of informational images—and is one that may influence the trajectory of nanotechnology as well as our understanding of atoms and our world. The scientific literacy this different narrative requires in order to read it involves not only learning how to read the story of nanotechnology through understanding scientific practices, but also, as reading the "IBM" images teaches, learning how we are also writing the story as we read.

"Nanospace:" Evocations of the Nanoworld

> *There is indeed, room at the bottom, and we are beginning to move in.*
>
> —John Brauman

Feynman employs another common scientific figuration in his speech: In describing the "room at the bottom," he comments, "It is a staggeringly small world that is below" (22). Feynman's expression of the room at the bottom as an entire world, albeit a "staggeringly small" one, efficiently and powerfully communicates to scientists and non-scientists alike, as the analogy conveys the importance of that room and suggests possibilities for how humans should understand the nanoscale. One reason that Feynman's evocation of a "staggeringly small" world is so effective is that Feynman creates a seemingly tidy analogy between the nanoscale and microscale through his reference to a scientific trope of small—often called subvisible—worlds. An association between what is seen with a microscope and possible views of other worlds have occurred in both text and image form starting from the earliest days of microscopy, and so forms an enduring scientific trope in microscopy. Feynman suggests that we can relate to nanoscale phenomena just as we relate to phenomena on the microscale—that is, according to the conventional trope, similarly to how we relate to phenomena on a macroscopic scale. John Brauman's suggestion that we are beginning to move into that room, in his introduction to Feynman's reprinted speech in a 1991 *Science* issue, makes explicit the human role that Feynman's analogy implies (1277).

Brauman's reference to Feynman is an indication of the role of the trope of subvisible worlds in articulations of nanotechnology. Indeed, the figuration of small, or subvisible, worlds appears in both textual and pictorial descriptions of nanoscale phenomena. For example, authors of scientific articles describe nanoscale data features in terms of terrain, such as "hills," "steps," and "terraces."[97] In another example, Binnig describes his reaction to seeing his and Rohrer's first "7x7 Silicon" image in terms of the subvisible worlds trope: "I could not stop looking at the images. It was like entering a new world" ("From Birth" 399).[98]

Images of nanoscale phenomena created by the STM and related visualization technologies also evoke landscapes and worlds through figurative means, using techniques such as false color and false perspective or shadowing. For example, the creators of an image of a copper surface have used false shadowing and false colors (coppers and browns) to indicate a rock-like surface against a light blue background. Evocations also occur in composite images, like the cover image of the National Science and Technology Council (NSTC) publication, "Nanotechnology: Shaping the World Atom by Atom," depicting a corrugated STM "planet" surface against a starry black sky, replete with a comet, an Earthlike planet, and a moon. STM researcher Shigeyuki Hosoki has even used the STM to create an image of letters that spell out "nanospace" seemingly in three dimensions, thereby articulating a place where one can not only read and write, but also into which one can enter (Grey 705).

In these textual and pictorial examples, on a general level, the trope of subvisible worlds functions to communicate information about nanotechnology by providing a frame for understanding the scale of atoms and small molecules. Moreover, the use of the subvisible world trope helps communicate a context for nanotechnology by articulating how humans can relate to nanoscale phenomena. For example, like Eigler and Schweizer's "IBM" images, the combination of writing and conventions of three-dimensional perspective in Hosoki's "nanospace" image allows viewers to draw analogies between the space depicted in Hosoki's image and the perspectival space with which viewers are familiar. Such analogies help viewers see themselves as not only becoming familiar with the nanoscale, but also "moving in" to inhabit the space. Additionally, an even closer look at the textual descriptions of new worlds in relation to nanotechnology shows that text articulations seem to fit into the conventional trope.

However, what the trope of subvisible worlds more specifically communicates about nanotechnology—and the role of humans in relation to nanotechnology—is less clear it than may seem at first glance. First, the comparison of the nanoscale to the macroscale is more tenuous than that of microscale to macroscale due to the nature of the differences between nanoscale phenomena and those on the macroscale. The differences between nanoscale and macroscale are significant; indeed, much of the promise of nanotechnology relies on the fact that atoms and molecules behave differently at the nanoscale than they do in larger masses at the micro- or macroscale. The laws of physics that explain human-scale events, including fundamental laws like the second law of thermodynamics, do not necessarily operate at the nanoscale (Chalmers). In fact, some researchers suggest that the transition between macroscale and nanoscale may constitute a major problem in actually using techniques that some tools seem to make possible, like atom manipulation for practical applications like molecular computers (Mulhall 40). As nanotechnology researcher Chad Mirkin, for example, observes, "the realization of many of these [nanotechnological] notions is hampered by a fundamental inability to wire up and interface structures on the nanometer-length scale with macroscopically addressable components" (185).

Such complications to what seems like an almost transparent set of related analogies suggest a closer look at how the analogies appear in STM images. Indeed, the shift in scale from nano- to macro-, STM imaging processes, and haptic vision practices suggest that the trope of subvisible worlds, too, may shift from the established conventional trope in ways that become important to how information about the nanoscale and our experiences with the nanoscale are conveyed in STM images and in discourses of nanotechnology. Like the trope of writing, the figuration of subvisible worlds, or nanoworlds, can also be read in terms of the conventional scientific trope. However, the trope can also be read in a way that takes into account the characteristics of the nanoscale that researchers engage with in order to create the images. As I'll elaborate below, the "nanoworld" trope also evokes a participatory "world" for the STM user and image viewer, a world that is primarily associated with computer-generated graphics and simulation programs, not microscopy. In associating the nanoscale with a participatory "world," the trope of subvisible nanoworlds sets the nanoscopic and macroscopic into an analogical relation that differs from that created by conventional microscopes so that viewers and readers may imagine themselves amidst nanoscale phenomena. The different relations the nanoworld trope creates thus also present different ways to understand the nanoscale as well as the promises and possibilities of nanotechnology.

Like the trope of writing, instantiations of the subvisible world trope in nanotechnology discourses highlight how concepts of the nanoworld and nanospace function rhetorically within depictions and descriptions of nanotechnology, and reveal the relations that are set up between viewer and viewed, or STM user/viewer and nanoscale phenomena. Reading the trope of subvisible worlds, in terms of the established scientific convention and how the trope is used in STM images, produces a view not only of the nanoscale, but also of ourselves as we approach atoms, interact with them, and literally as well as figuratively assist in producing the nanoscale and nanotechnology. Therefore, reading the trope of subvisible worlds can elucidate how we conceive of nanoscale objects of scientific knowledge, and can also clarify our relations to those objects as well as to the nanoscale.

From Microworld to Nanoworld? The Conventional Trope of Subvisible Worlds

One way of reading researchers' and proponents' recurrent evocations of landscapes or worlds of nanotechnology is in the context of a common trope in microscopy since at least 1631 in the Netherlands and in England since the 1665 publication of Robert Hooke's popular *Micrographia* (Alpers 7). Like those creating evocations of the nanoworld, Hooke was engaged in the task of presenting information about what had not been seen before as well as what information a new instrument could produce: In the mid-seventeenth century, the microscope was still a new instrument, having emerged in Europe around 1610, at the same time as the telescope. Hooke uses the trope of new worlds in *Micrographia*, such as in a commonly quoted example in the fourth page of the preface: "there is a new visible World discovered to the understanding." Hooke also uses pictorial landscape conventions to frame some of his drawings of microscopic observations, such as an image of blue mold (plate opposite 125). The two separated views presented in this image resemble landscape drawings in that the first depicts what seem to be leafless flowers and buds growing out of a grassy substance against a dark background, while the other view resembles map drawings of islands. Additionally, the views of blue mold also include a slightly angled depth perspective, much like that used in Hosoki's image of the word "nanospace" or in the color versions of the "IBM" image (Figure 2).

Hooke's image evokes a connection to a landscape or another world that communicates two intertwined ideas about microscopic phenomena and the promise of microscopic exploration. First, the connection to a landscape or world allows microscope users to explain the less imaginable—the microscopic—in terms of the imaginable. The trope of subvisible worlds helps render the "landscape" as a recognizable given, a backdrop against which objects may appear. For example, early microscopists linked their explanations of microscopical phenomena to a concept whose existence, Anne-Julia Zwierlein argues, was anticipated in fiction (69). Using a previously established trope of subvisible worlds to explain microscopical phenomena lends Hooke's (and other microscopists') observations credibility, as Hooke's viewers could then understand the strange in terms of familiar forms. (Of course, Hooke's use of the trope also most likely shaped what viewers expected to see.) Microscopists' observations, and the analogy to a new world, gained credibility through the association with Bacon's conception of the scientist as an explorer searching for new land (see, for example, sec. CXIII and sec. CXIV).

Second, the trope of subvisible worlds helped Hooke and other early microscopists explain what microscope users *should* see, an important job considering that anyone who did use a microscope to see mold similar to Hooke's example would discover a large difference between Hooke's image and what the microscope user saw.[99] Early microscopes frustrated any idea of transparent viewing, partly because optical illusions and distortions form an integral part of the viewing experience of the instrument, but also because viewers cannot confirm with their eyes that what they see with a microscope exists. The trope of subvisible worlds, then, allows microscope image viewers to imagine similarities between the world viewers know and what the microscope images describe, allowing viewers to imagine that they are navigating microscopic space. The trope helps microscopists gain credibility for their work while also addressing the questions of authenticity that have always accompanied descriptions of objects unverifiable by the unaided eye.

The conventional trope of subvisible worlds also helps communicate what has been learned about microscopic phenomena, as it implies what is still left to explore, evoking wonder and excitement about what is already discovered and what is left to discover. In framing the images of microscopical observations in terms of part of a world instead of something more bounded, like an object, microscopists such as Hooke suggest that there is much more to discover. Additionally, the analogy of subvisible worlds communicates that discovery with a microscope is not only an intellectual one; the analogy describes a bodily relation between the subvisible world and the viewer. Through the suggestion of a possible new world, the trope also then implies that viewers might experience the worlds themselves by turning to the microscope. As philosopher of science Ian Hacking explains, in using the figuration to explain how we learn to use microscopes: "In short, we learn to move around in the microscopic world . . . we enter the new worlds within worlds that the microscope reveals to us" (209). The specific details of the microscope user's bodily relation are not described in this analogy, but are instead left up to the viewer or reader to imagine. In this, some limits of the analogy appear: While the viewer of microscale landscapes can imagine himself or herself as an explorer of new worlds, the credibility of what is presented in microscopical images also depends on objective presentation, so that microscale images present new worlds in a pristine, untouched state. The viewer does not—and cannot—go to the new worlds, for the images must remain objective. Instead, the viewer is invited to marvel at the vistas of the subvisible world. As Barbara Maria Stafford and Keller, in "The Biological Gaze," explain, while there is always some manipulation of samples in order to actually use microscopes, microscopists de-emphasized the importance or impact of their manipulations of the samples, for what was true was to be conveyed objectively, without bodily action, and only to the eye.

Seeing the Nanoworld: Tropes and Scanning Tunneling Microscopes

At first, the conventional trope of subvisible worlds seems directly aligned with references to new worlds or small worlds in visual and verbal descriptions of nanoscale phenomena from researchers. For example, Binnig and Roher's statement that seeing the "Silicon 7x7" images was like "entering a new world" ("From Birth" 399) can be read in the tradition of Hooke's description of a "new visible World discovered to the understanding" on the fourth page of his preface. The conventional trope also seems to function in images: For example, in Hosoki's nanospace, the nanoworld is portrayed as the assumed, steady surface or medium in which object-like atoms or molecules exist. Indeed, even the raw data collected by the STM looks topographical, since it appears on a computer screen in a spatial matrix pattern, regardless of whether the STM is presenting topographic or electronic data. The visual format of the data thus lends itself to being treated as a landscape, even in the formation of images. As Dominique Brodbeck, who created a variation of the "Quantum Corral" image, observes, "From an image creator's point of view, what I always liked about working with STM data was the fact that the surfaces are very similar to landscapes, and that you can apply the same design guidelines and intuition as you do in landscape photography" (qtd. in Frankel 261).

On closer inspection, the trope of subvisible worlds loses some of its expressive power because the trope stops communicating the new or unique knowledge of the nanoscale that researchers have claimed is what makes the nanoscale exciting. Two related STM images that use the landscape metaphor illustrate how the conventional reading falls short of expressing what STM images convey. The first image appears in M. F. Crommie, C. P. Lutz, and Eigler's *Science* article, "Confinement of Electrons to Quantum Corrals on a Metal Surface." In the article, Crommie, Lutz, and Eigler report how they arranged iron atoms in a circle to form atomic corrals, thereby creating a space that traps electrons in order to observe the behavior of the trapped electrons. The article includes two images of the same "corral": a "top view" of the corral and another image whose perspective is positioned at a slight angle so that viewers see the depth of the concentric, ripple-like "wave pattern caused by the interference of incident and scattered surface state electrons" (218) emanating from the center of the image, much like water after a pebble has been thrown, emphasized by the play of light and shadow. The corral structure is also imaged in the "Quantum Corral" image discussed earlier (Figure 4) that appears on the IBM website. The second image positions viewers to peer between red and blue pointed atom peaks across a rippled floor, towards a far row of atoms set against a stark black background.[100]

Although the "corral" images convey a view of a landscape in a subvisible world, the nanoscale "view" communicates more than the idea of a new world. Text accompanying the corral image on the IBM website points to an additional message: "The discovery of the STM's ability to image variations in the density distribution of surface state electrons created in the artists a compulsion to have complete control of not only the atomic landscape, but the electronic landscape also" ("Corral Reef"). Here, the "artists" (i.e., the scientists who created the corral image) point out quite explicitly how they have not discovered a nanoscale "landscape," but instead how the scientists created it.

While the use of landscape in corral images perhaps, at first, suggests a three-dimensional landscape similar to the natural, human-scale world, the image works against the convention, especially through the emphasis on the user's hand in creating the space of the nanoscale. This emphasis on explicit creation is further communicated in particular through the researchers' mention of "complete control."

If the analogy of subvisible worlds is read conventionally, the emphasis on interaction at the nanoscale—and the influence that interaction has on understanding nanotechnology—is not communicated. The association between nanoworld and microworld suggested by a reading of the trope of subvisible worlds, in terms of the conventional microscope version of the trope, obscures the changes that accompany shifts in scale. Because of the different behaviors of phenomena at the nanoscale, shifts in scale are more difficult to bridge in reality than in image or language. Therefore, the associations created by the trope of subvisible worlds in microscopes obscures the nanoscale's unique characteristics, also suggesting an easy, but less than accurate, comparison with the microworld and, through this, the world of the human scale. The conventional microscope figuration also hides the participation and role of humans in creating these images—and, in the case of the quantum corral, the conventional reading of the trope also hides the value of the portrayed landscape. Therefore, reading the trope conventionally only communicates the idea of a new world to explore. While the established frame of linking the nanoscopic to microscopic and macroscopic scales may seem useful in communicating the excitement of a new field to non-scientists, the conventional trope does not fully express the scientific meaning communicated within the images.

Nanospace as Digital Space: Computer-Generated Worlds

The processes by which the nanoscale is made visible suggest another element that may serve as a more relevant vehicle in the spatial metaphors of STM images, one that emphasizes the manipulation inherent in visualizing the nanoscale while also playing down the simple analogy to the "macro" world. Hosoki's "nanospace" image, for example, features this new element. While Hosoki's image can be read in terms of microscopists' conventional subvisible worlds, it also evokes another connotation for contemporary viewers that competes with the trope of subvisible worlds: The surface depicted is familiar, as James Elkins remarks, "very like the surface of a table" (Object 60), echoing another familiar computer-visualized table surface—the desktop. The association of Hosoki's image with a computer-generated interface that has trained so many in first-world Western cultures to interact with images as they compute is particularly rich in that the link articulates not only a visual resemblance to a human-made object, but also structures a relation with the viewer in which she or he is solicited to respond. A viewer does not tend to only look at a desktop image; instead, the image functions as a behavioral cue (Drucker "Reading" 215). Therefore, a viewer interacts with the image by using the image as an interface with the hardware and software. The association of the space within Hosoki's image with that of a desktop image also alludes to the interactions that help to compose the STM image.

While microscope users have always engaged in some hapticity as they manipulate specimen samples for viewing, as Keller argues in "The Biological Gaze," the STM apparatus and image production processes intensify the engagement of microscope users with the sampled surfaces so that what emerges is a co-created nanoworld. The "corral" images, for example, argue for a manipulable nanoworld: What the corral images show is not a "landscape" that STM users have happened upon, but a human-made artifact, a "corral" created by the researchers' manipulation of iron atoms into a circle. The space the corral images articulate could be more accurately compared with a coliseum than a landscape in that the corral was built to test the behaviors of atoms placed within it, and was built using the same participatory practices that Eigler and Schweizer used to write "IBM." Crommie, Lutz, and Eigler use the imaging process as part of the experiment construction, not only as part of the experimental results (218). Indeed, the word "corral" does not entirely fit the description of the image: The low angle of the view in the "corral" image highlights the grandness and accomplishment of the structure itself, creating a focus on the spectacle of the building as well as the wave functions within.[101]

The explicit focus on researchers' construction of the "nanospace" found within the experiment design and STM images suggest a different referent than the macroscopic world for the trope of subvisible worlds found in the corral image; instead, the world referred to links to a computer-generated world, one enabled by the properties of the computer affording interaction with images as interfaces. As previous chapters explore, interaction creates different effects on viewers and image-creators. One effect of using computers for imaging (and the ability of computer users to alter images they create almost instantly) is that, as Elkins suggests, how we conceptualize space has changed. Writing about the tendency of artists and illustrators to create simpler, less complicated constructions, he observes: "The lumpy, crowded spaces of Western painting have been replaced by the sheer, limitless spaces of contemporary graphics" ("Art History" 335). The plane on which the corral rests, an expanse evoked by a dark, far horizon, exemplifies this vast space.

The computer-generated characteristic of limitless space influences the nanoworld trope so that, on occasion, viewers find themselves moving over a seemingly vast scape in images such as the corral images, especially given the viewer's position at one end of the structure (see Figure 4). Eigler's comment about his response to the color quantum corral image he created points to the nature of this experience:

> These things [the standing waves in the corral's interior] were entirely predictable, and we routinely work them out in our problem sets when we take courses in quantum mechanics. We knew of them in a purely cerebral way. But here they are, alive to our eyes and responsive to our hands . . . [ellipses in original] quantum mechanics made visceral! The realization that you can build a quantum state of your own design, see the result, and change and shape the quantum state like a lump of clay . . . well that's just candy to a physicist. (qtd. in Frankel 261)

The limitless, manipulable, space that the trope evokes figuratively magnifies the nanoscale, helping to create what Eigler calls "a delectable intimacy between us and the quantum world" (qtd. in Frankel 261). The relation implied in Eigler's comments between human and physical phenomena is not the same as Hooke's "new visible world discovered to the understanding"—implied instead is that knowledge of the quantum world exists in the ability to interact within nanospace. The computer-generated characteristic of limitless space thus presents an alternate analogy for expressing knowledge of the nanoscale, an analogy that emphasizes embodied interaction.

The association of nanospace with computer space implies that the relation between nanospace and viewer is unlike exploration of other worlds; instead, the use of the trope in STM images aligns nano-landscapes with other computer-generated images such as those produced by virtual reality programs, whose user also moves across or over a data matrix, allowing the anticipation of an endless participation between pixels and person. The ability to adapt the image also allows image creators and viewers to participate in conventions of haptic image-interfaces, such as those described in Chapter 3, and so also develop practices of anticipating that the image may change in part due to viewers' responses. Anticipatory practices encourage viewers to envision many possibilities for exploration, intensifying viewers' sense of immersion and contact with what the image shows.

However, association with computer-generated imaging raises questions about scientific images in that issues of authenticity and credibility occur even more insistently than with traditional microscope images. While traditional microscopists have answered the charge of authenticity and credibility in part by relating the "new visible world" to the macroscopic world, this strategy does not work as surely when images—and atoms—can be manipulated. Making STM images overtly computer-generated, and so linking them to the limitless spaces associated with computer graphics, offers one solution to the question of authenticity as well as to the problem of shifting scales through the articulation of what the computer creates as a world.

The tendency of many, including scientists, to think of what the computer can create as an independent world lends weight to the change in the trope of subvisible worlds. As Galison explains, during and after World War II, physicists and engineers began using computers to create simulations that they viewed as "an alternative reality" in which they could conduct complete experiments in designing the hydrogen bomb ("Computer" 119–120). Simulations, considered self-sufficient worlds within the machine, are what make the computer a pre-eminent tool, as Paul Edwards explains in his history of the computer in relation to the Cold War (171–172). Further describing the pull of computer-generated simulations as worlds, Galison argues that the computer-generated world, "even in its earliest days, held enough structure on its own to captivate its practitioners" ("Computer" 154). Simulations allow for consistent events to happen within the parameters of the simulations (their worlds), and do not make a transition between the internal world and the external world; these characteristics of simulations led researchers to create artificial life. Furthermore, as anthropologist Stefan Helmreich documents, from the 1970s on, computer programmers began calling simulations "microworlds," perhaps influenced by, as one of Helmreich's scientist informants suggests, physicists' views of quantum mechanics (75–76). Among the researchers of artificial life he studied, Helmreich observes: "This notion that computers can contain separate, closed worlds came up again and again, and was one of the rhetorical moves people used to grant simulations their own ontology" (78). The linking of worlds that Helmreich observed, he argues, became stronger as the notion of a "microworld" shifted to "artificial world," emphasizing the self-contained independence of the world, as opposed to a shift in scale (71). The link to computer-generated worlds, then, does not allow for an easy shift from nanoscale to macroscale phenomena. The link to simulations also gives the trope of subvisible worlds in nanotechnology another possible benefit—an association with the predictability of simulation or (relative) lack of complexity—thus encouraging viewers about the possibility of understanding the complex phenomena of the nanoscale.

In addition to the connection with computer worlds expressed in the use of the subvisible worlds trope in STM images, the version of the nanoworld as a separate, contained world aligns with the "microworlds" that twentieth (and presumably twenty-first) century physicists have relied on, and helps elucidate how the trope of nanospace or the nanoworld functions. Jed Z. Buchwald relates how physicists use the existence of a microworld in their microphysical practices. Buchwald points out that the trope of microworlds operates as a point of commonality among physicists who use different methods of gathering data about atoms (such as the STM and atom collider) (210). Buchwald argues that such a practice originates in nineteenth-century chemistry and physics, recounting that the microworld became "real" in the 1890s, as it emerged from the concept of ether (210). The concept of worlds became important in articulating knowledge about what cannot be seen, allowing physicists to discuss and spatially visualize atoms.

In tapping into associations with computer simulations and the trope of the microworld in physics, as well as into the experiences of both scientists and non-scientists with the computer, the trope of a subvisible nanoworld—whose vehicle is a computer-generated world—accounts for the gulf between the nanoscale and the macroscale by positing a separate world at the nanoscale. Like those created through the conventional microscope trope, the nanoscale "world" allows researchers to imagine themselves within it. However, the nanoworld also allows scientists to create structures within the nanoworld as they explore. Therefore, STM images that depict nanoworlds not only document an area that has been explored, but also forms interfaces for exploration for the microscope user as well as for the viewer of images of the nanoscale created by the microscope.

Building the Nanoworld

In the movement from micro- to nano-, the trope of subvisible worlds used to articulate the nanoscale seems to align less with the conventional microscopical evocation, and instead seems to evoke connotations to simulated microworlds or computer-generated virtual worlds. While the trope of subvisible worlds has always conveyed some sense of bodily relation with the objects under view, the STM user's participation in creating the image allows the nanoscale to be explored not only through viewing but also, more importantly, through manipulation. The combination of the imaging process of the STM and image production process through graphics programs to highlight certain data emphasizes this exploration: Images of the nanoworld are not only artifacts (like photographs snapped of a foreign land); they are explicitly tools with which experimenters, and possibly viewers, can build that land as they experiment, much like virtual worlds such as *Second Life*.[102] The use of STM images as tools also positions haptic vision as a method for scientific discovery. As Eigler explains,

> This ability ["to fabricate on the atomic scale"] can best be viewed as a tool which the scientist can use to explore intellectual *terra incognita*. It is the excitement that comes from this exploration that ultimately is the real driving force for building things with atoms today. ("From the Bottom Up" 427)

Like writing, it is the action or process of building that Eigler suggests generates excitement about the nanoscale.

The emphasis on experiment, leaning on the possibilities of computer-generated images for interaction, changes the expression of the trope of subvisible worlds in STM images. Instead of focusing on recognition—the new visible world that is similar to our world, as Hooke's images portray—the trope shifts focus to a world whose limitless vistas and invitation to the process of imaging encourages its viewers to explore. Here, the very artificiality of the nano-landscapes—the bright colors and structures, for example, as well as the image planes stretching into the distance—emphasize the difference in associations. The nanoworld trope, like that of writing, becomes more closely associated with an action. The nanoworld trope also expresses a relation between nanoworld and macroworld that does not rely on analogy per se, like the conventional microscope trope; instead, the nanoworld relation depends on analogous *actions*—responses to the responses the STM researchers receive from the nanoscale and the STM apparatus through the image interface. The focus on analogous actions thus highlights the work of the human hand in also forming the nanoworld, reinforcing a sense of knowledge as not only accessible, but also manipulable and responsive.

The shift in trope documented above could be understood as merely another reference to the wider cultural presence of the computer that engages viewers by making fleeting, perhaps decorative, connections to the familiar—or to the fun, in relation to computer games. However, the shift in trope communicates something significant about science as well: a key process of knowledge-making in nanotechnology, exemplified in the interactions necessary to visualize atoms using the STM. The rhetorical figure of nanoworlds thus indicates an integral articulation of the nanoscale. Indeed, the nanoscale that the nanoworld trope articulates suggests that both the content and the form of the images of nanotechnology that express the trope are significant for understanding how knowledge of the nanoscale shapes the known and the knower, as well as for communication of and responses to both that knowledge and our relation to knowledge.

The shift in trope to creating associations to computer-generated worlds highlights a way of making knowledge that differs from that found in conventional microscopy, a way that relies on the expanded functions of computer-generated images as a tool for experiment. More specifically, the shifted trope of nanoworld expresses the experience of co-creating the nanoworld as similar to creating or altering a digital image and, through the image, to creating or altering a computer-generated landscape or world. Thus, the trope communicates the experience or event of the experiment not only by presenting evidence of nanoscale phenomena and of the experiment, but also by inviting viewers of the images to participate, whether those viewers alter the image themselves or if viewers imagine themselves in the space of the experiment. Just as making nanoscale phenomena visible requires interaction, so does making the tropes' association: the images solicit viewer interaction through associations with computer desktops, video games, or other virtual worlds. The trope of nanoworlds also creates associations between STM images and other experiential, computer-generated modes of learning about the nanoscale, such as video games, that Colin Milburn discusses in relation to nanotechnology (see "Atoms," "Digital"). The changed trope of subvisible worlds thus also presents a strong case of Black's interaction theory of how metaphors work, in that as we view the nanoscale as a nanoworld associated with a computer-generated world via the digital image, we see the nanoscale as a space in which to interact, albeit virtually.

Like the "IBM" image, images of nanospace slip into discourses other than those surrounding and running through scientific papers or journal covers on which the images might first appear. For example, images of the quantum corral have won the "cover-page 'triple crown,'" as physicist Eric Heller puts it (489), appearing on the *Science* issue in which the article appears, as well as *Nature* (volume 363) and *Physics Today* (November 1993). Besides reaching a general scientist audience, "Quantum Corral" also appears on the cover of B. C. Crandall's edited collection of predictions, *Nanotechnology: Speculations on Molecular Abundance.* A less colorful image of the corral appears on *Nanotechnology*, edited by Gregory Timp, in the U.S. National Science and Technology Council (NSTC) brochure "Nanotechnology: Shaping the World Atom by Atom" that is "intended for public use" and appears on websites. These images reach a non-scientific (or non-expert) audience as a representative image of nanotechnology. In fact, the NSTC brochure contains two other images that evoke the nanoworld and a nano-landscape, reinforcing the trope for the public audience of the brochure.

Figurations like the nanoworld trope help to establish a vision of what the nanoscale may *look* like through what the nanoscale may *feel* like, regardless of the realities of light wave size, the different quantum behaviors of individual atoms and electrons, and any protestations that STM images are not really what the nanoscale "looks like," as happens often with quantum physics images (Elkins, "Logic" 164). In so doing, figurations such as the nanoworld analogy establish the boundaries of the room at the bottom that Brauman asserts we have begun moving into, and so help frame how we not only "see" this other world from afar, but also how we relate to and interact with nanoscale phenomena. What a reading of the trope of the nanoworld, in terms of the capabilities of the STM apparatus and the functions of STM image, reveals is that STM images expressing the nanoworld reflect something of the STM user in them: As the trope of the nanoworld emphasizes manipulation, the expressions of the trope also show the impact of the human on creating the nanoscale. The explicit human imprint on the nanoscale found in images like the corral images allow us to create different relations with the nanoscale, relations in which what has changed is not only the scale, but ourselves as we change in our interactions with these informational images and with atoms. Images themselves become part of the experiment as we co-produce a world we discover as we help create the nanoscale. What becomes visible, too, is not only a different relation to the "room at the bottom," but also the atom we can touch and experiment with—the emergence of the atom into a nanoworld that we not only "see" but also control as we co-produce.

The visual components of digital images that constitute the scientific images of the nanoscale, and that here constitute a seeming allusion to a conventional microscope trope, also form part of the persuasive language of nanotechnology, as the components not only help contextualize how we see the nanoscale, but also how we envision ourselves amidst nanoscale phenomena. As discussed above, the association of the nanoworld with a separate, closed world of computer-generated simulation helps readers conceive of the nanoworld as an authentic space that is not exactly like ours, thus helping researchers overcome questions of credibility and of shifting scales from nanoworld to macroworld.

The nanoworld trope also positions the nanoscopic and macroscopic in a different analogical relation than that of the microscopic and macroscopic. Unlike microscopic worlds, whose vistas are depicted in early drawings of microscopical views, the nanoworld can be virtually entered—constructed, even—calling forth haptic and visual responses from viewers. The articulation of the nanoscale as nanoworld creates a relation between nanoscale and macroscale that depends on responses as part of the interactive process, for example: electron responses to other electrons used to gather data about nanoscale phenomena that compose the image; responses of researchers such as Crommie, Lutz, and Eigler, who created the corral; and responses of researchers to create the corral image so that the image includes elements that create the trope of the nanoworld. The nanoworld thus becomes framed as a series of events in which the user or viewer is involved, reinforcing the idea of the nanoscale as not only accessible, but also manipulable and responsive. Further, the framing of the nanoworld as manipulable suggests possibilities for envisioning the larger world as manipulable and responsive, aligning with Alfred Nordmann's observation about the title of the brochure "Nanotechnology: Shaping the World Atom by Atom": "it is no longer assumed that 'the world' is a given, and as such, subject to representation on the one hand [in science], technological intervention on the other [in engineering]. Instead, nanotechnology is constantly shaping and reshaping the world" (52). As we have started to enter the "room at the bottom," we have found plenty to see and touch; but in that experience and process, we also have been touched. As we visualize the nanoscale, we participate in a world we discover as we help create it. We cannot help but touch and be touched—our interaction helps call our vision of the nanoscale into being. Through the trope of the nanoworld, what becomes visible is expression of a responsive relation not only to that room at the bottom, but also to ourselves and the atoms we touch—the nanoworld we co-produce.

Conclusion: Visual Intelligence

Feynman's speech, "There's Plenty of Room at the Bottom," articulates one of the promises of nanotechnology currently used to promote the field: building machines from molecules or atoms. Feynman speculates about a coming field that would focus on "the problem of manipulating and controlling things on a small scale" (22). It is interesting that the small scale about which Feynman speaks becomes visible only through interaction as well—interaction with the computer, with visualization technologies such as the scanning tunneling microscope, and with the image as well as the atom. The participation of human hands in building the nanoscale becomes visible through the tropes of writing and subvisible worlds found in the images discussed above.

Analyzing instantiations of writing and nanoworld tropes produces an understanding, a sort of visual intelligence, about how STM images may not only communicate science, but also communicate *about* science: STM images help argue for the believability of the objects the images communicate about, not only through their data presentations, but also through their visual metaphors and other figurations. The process of making atoms visible has altered the expression of these two common scientific tropes found within the STM images discussed above, so that the tropes no longer communicate within the bounds of conventional uses of the tropes. Instead, the images create new associations to communicate knowledge about the nanoscale. In so doing, following Black's interaction theory of metaphor, new associations the images create in turn affect the shaping of knowledge of the nanoscale.

Emphasis on interaction expressed in STM images provides versions of the tropes of writing and subvisible worlds that reinforce the sense of knowledge about the nanoscale as manipulable and responsive that, in turn, affects how knowledge of the nanoscale is shaped. The tropes of writing and nanoworld in STM images articulate the nanoscale as a participatory space in which knowledge is not gained by just looking, but by interacting. In order to see as well as make sense of nanoscale phenomena, one must interact. Visualization practices affect scientific knowledge production: As expressions in STM images of the tropes of writing and nanoworld attest, researchers visually—haptically—co-create the nanoworld as they see atoms.

5 Conclusion

Haptic Visions of Science and Rhetoric: Interaction and Its Implications

Richard Feynman's 1959 talk "There's Plenty of Room at the Bottom," and the images that D. M. Eigler and E. K. Schweizer published in *Nature* serve as touchstones for proponents of nanotechnology not only for the use of common scientific metaphors of tiny writing or landscapes discussed in Chapter 4, but also for references to another vision of the future that specifically includes interaction: of one day building structures from atoms, from the "bottom up." The vision of bottom-up manufacturing also appears in many current articulations of nanotechnology. For example, manufacturing from atoms on up forms one of the four "signature initiatives" developed in 2011 by the National Nanotechnology Initiative (NNI), the umbrella organization for U.S. government investment and direction in nanotechnology. The goal of the signature initiative for "sustainable nanomanufacturing" is to create flexible, "'bottom-up" or "'top-down/bottom-up' continuous assembly methods to construct elaborate systems of complex nanodevices" (United States, Subcommittee, Supplement to the 2014 Budget 51). The NNI has supported the initiative verbally and financially: $23 million of the $1.76 billion requested for the 2011 budget was allocated to this initiative (United States, Subcommittee, Supplement to the 2011 Budget 6); in 2014, $60 million is dedicated to nanomanufacturing (United States, Subcommittee, Supplement to the 2014 Budget 10).

While the goal of the NNI nanomanipulation initiative is to move beyond demonstrations of individual atom manipulation found in the STM images discussed in this book, the NNI initiative relies on the same understanding of atoms as tangible and manipulable that STM images have helped to communicate to scientists and non-scientists alike. As this book explores, the perception of atoms as tangible and manipulable is supported by disparate, yet linked practices that emerge out of historical and cultural traditions. These practices, ranging from STM operating dynamics, to imaging processes, and to visual conventions, synergistically create a field of visibility in which researchers can see and manipulate atoms and other nanoscale phenomena. However, as this book has also discussed, a closer look at the STM imaging practices associated with making atoms visible reveals that STM images do not only support the promise of the ability of nanotechnology to build structures from the bottom up; the necessary, increased interactions with the nanoscale and with images also resonate with—and help change—other aspects of scientific practice and communication. Thus, *Haptic Visions* also reveals a complex story about changing practices in imaging and the production of scientific knowledge. The changes in imaging and scientific knowledge production suggest ways that rhetorical knowledge and practice are also changing, or may change as we communicate through digital images such as those produced with the STM. What follows sketches some of the implications of the increased interactions necessary for visualizing the nanoscale with the STM—implications for imaging, the production of science and our place within such productions, and for rhetorical knowledge and practice.

Visions of Nanotechnology, Visions of Science

As this book has described, the process of making atoms visible that researchers using the STM undertake while creating images has shaped the presentation of data in image form. The image-creation processes that STM researchers use provides a good example of how information and image form affect each other, and so influence understandings of what an image expresses—here, understandings of the nanoscale. The increased interactions at all levels, from atoms on up, that are inherent in STM imaging processes, and that allow for the nanoscale to become visible, expand the functions of the image to include not only documentation of experimental evidence, but also (in some cases) the formation of the experimental apparatus. The image serves as an interface for the experimenter to interact with nanoscale phenomena, and so the digital image becomes an experimental tool as well as communicative medium.

Thus, the event of the image creation *becomes* the experiment while also serving as a record of that experiment.

The focus on interaction that STM images emphasize raises questions for further research about the effects of interaction on knowledge-making and communication practices, questions that reveal outlines of larger issues that are sparked by the connected changes in visualization technologies, imaging, and relation to nanoscale phenomena. If, for example, as the shifts in scientific practices recounted in *Haptic Visions* suggest, knowledge-making becomes more explicitly participatory, what responsibilities for communicating and for responding to participants—including nanoscale phenomena as well as those who view the images—does this emphasis on participation create for researchers? How are researchers also changing as the researchers respond to the nanoscale while they co-create that nanoscale? While STM images express the nanoscale as a co-created, participatory space, tracing the functions of STM images as evidence or as experiment influence or inform current discussions of who should or can participate—and how—in the development of nanotechnology would be a productive direction to take. Further investigating the details of how STM images create and then extend rhetorics of participation to viewers of images who are not microscope users in ways that either invite or forestall actual participation in the making of science is also productive in tracing the impact of haptic vision practices.

Re-articulations of the relations between nanoscale and macroscale, and between nanoscale phenomena and researchers, through expanded interactive functions of the image created by shifted vision practices, visual conventions, and tropes, also participate in a wider narrative of science. For example, the intensified interactions necessary to visualize the nanoscale also connect to changes in understanding the shape of science in regard to researchers' relations to science. Lorraine Daston and Peter Galison posit that nanotechnology departs from ideas of "pure" science (395), changing the concept of the scientific self and creating a hybrid engineering-scientific self (414). The "Quantum Corral" (Figure 4), for example, seems to fit within Daston and Galison's view, as the corral is created as part of the experiment, and so presents evidence of "engineering" the nanoscale. How hybrid engineering-scientific selves shape, and yet are also shaped by, nanoscale phenomena as scientists produce and communicate knowledge are questions that become important to understanding science. Philosopher Karen Barad proposes one way to articulate participatory aspects of making scientific knowledge in her framework of "agential realism," as the framework takes into account roles of "human and nonhuman, material and discursive, and natural and cultural factors in science and other social-material practices" (26). Furthermore, she argues that our interactions with nanoscale phenomena provide an occasion to observe the deep entanglements that exist in all of our relations with the world. The view of a changed science that Daston and Galison and Barad present aligns in interesting ways with how some scientific images, such as STM images, communicate. The image of science that shifted practices of science collectively present is one in which data that is "informed" by its form becomes an interface for interacting with the world. The capacity of digital scientific and medical images to function as an interface allows for interaction in STM images with the nanoscale phenomena that are made visible in STM images; however, the interface function may also be a way to understand interaction with other phenomena, or even other kinds of data, as the allusion to computer-generated worlds discussed in Chapter 4 suggests. What becomes visible, then, in STM images is not only nanoscale phenomena themselves, but also the imbrication of the nanoscale with our knowledge structures and ourselves.[103] What my analysis shows is that the focus on interaction through the image-interface not only helps to shape nanotechnology into an "information science," but also changes the ways in which we interact with data and with the nanoscale *through* the form of the image-interface. What becomes visible in STM images, then, is not only data and the nanoscale, but also the interactive interrelationships between data, our knowledge structures, and ourselves necessary to visualize the nanoscale.

Visions of Rhetoric: Changes in Rhetorical Knowledge and Practice

The STM imaging practices that provide a glimpse into broader changes in science also have implications for rhetorical knowledge and practice. As this book has shown, attending to imaging processes when analyzing the rhetorical work of an image can reveal aspects of rhetorical elements that frame both image-creators' and image-viewers' experiences of the information the images communicate. Analyzing images through considering imaging processes thus highlights the complex work images do in not only conveying information, but also constituting how and of what we are persuaded.

The interactive production and viewing practices necessary to the workings of the STM, and the production of data in image form, affect the visual rhetorics of STM images. Analysis of the production and vision practices described in the first two chapters reveals the rhetorical capacities of STM images that are not entirely new. Yet, the rhetorical capacities of STM images present different characteristics than non-digital images due to the intensities of engagement found in some of the practices. For example, while researchers have always arranged data in some way in images, whether by cropping a photograph or applying false color, the intensified time and processing needed to create STM images from the beginning of the experiment through the final appearance of the image in a journal article emphasize different aspects of the image and, just as important, the intensified time and processing affordances emphasize the experience of the image as an event, as opposed to a final product. As Chapter 2 relates, STM-using and STM image-viewing experiences also create a different relation between image-maker, reader, and object of knowledge, disrupting the rhetorical situation often called the Aristotelian triangle so that all three—speaker, message, and audience—participate in a way that can be described as ontogenetic. Experiential, affective, and representational aspects of images function rhetorically as readers engage in the viewing/reading process needed to understand STM images, as Chapter 3 explains. Chapter 4 analyzed rhetorical registers of tropes in the images that express the experiential focus of STM image rhetorics.

The emphasis on interaction in the image form also shows in the conventions—rhetorical and otherwise—used to communicate about the nanoscale. STM images such as those discussed in this book communicate in ways that obscure and reveal what it is that the images show, in part depending on assumptions or viewing conventions that viewers bring to the reading process. The possibility of multiple readings suggests that the images also mark a moment in which old conventions are being broken. The possibility of multiple readings also indicates that new conventions are forming, such as the emphasis on the computer-generated nature of the images even when using perspective, and the repositioning of viewer and object to lead viewers through the data.

Analysis and exploration of conventional change expands the focus of visual rhetoric on intertextuality, "the recognition and referencing of images from one scene to another," through studying how conventions, tropes, and even vision practices can also serve as points of recognition and reference, even across media (Hill and Helmers 5). Awareness of intertextuality is critical to understanding the content of the images. As Charles A. Hill and Marguerite Helmers point out, for example, "if the reader is unaware of the precursors, the image will have a different meaning or no meaning at all" (5). The non-representational and representational aspects of images play parts in analysis of conventions, tropes, and vision practices and, as this book argues, should be considered in rhetorical analysis. While the dominance of perspectival vision practices helps explain why we keep returning to representation in visual rhetoric through expectations of how we make meaning, exploring haptic vision practices helps to focus more directly on what data (as opposed to objects) might look like, in addition to how elements inherent in the formation and circulation of images might form persuasive rhetorical dynamics. We could ask ourselves, for example, as we analyze a STM image or other digital informational image, what counts as rhetoric in the moments of immersion critical to haptic vision? Further research into rhetorics inherent in haptic vision practices in other contexts could explore concepts of entanglement, following Barad's proposal for understanding our place in the world, and ask, what are the ways in which communication emerges within a model of entanglement? Studying how other images made in other media, for other purposes, also summon experiential interactions in which persuasion and meaning-making occurs within the interaction or creation, not necessarily in the moment of "transmission," would continue to develop an expanded understanding of what counts as rhetorical knowledge.

In addition, emphasis on the image as interface, and consequent emphasis on response and embodied experiences of practice, also expands rhetorical practice. While Collin Gifford Brooke observes that emphasis on the interface is a hallmark of new media, my analysis demonstrates that emphasis on the interface does not only have an aesthetic or communicative register, and is not restricted to what is usually categorized as new media—art or literature or film, for example (xiii). The interface that the STM image comprises is also deeply imbricated within scientific practices of knowledge creation and communication. *Haptic Visions* shows another instance, beyond categories usually studied by new media or digital media theorists, of how form and information are co-created in ways that affect the formation of cultural knowledge—in this case, scientific knowledge. Rhetoricians should then not only study interfaces as objects of analysis, for example, but also consider how the changing practices associated with the processes of creating and using interfaces, such as the images presented in this study, reveal rhetorical practices that are themselves altering. Studying alterations of rhetorical practices highlights how what we consider objects of rhetorical study may also be changing.

In this study, the interactions between atomic phenomena, image, image creator, and image viewers also create a heightened sense of interdependence as part of the communication process. In the case of atom manipulation, interdependence suggests not a situation of mastery and control, but of participation based on response that then affects the communication process. The communication that produces the STM image, for example, is not one based on Claude Shannon's model of communication (as occurring between sender and receiver), but is instead one closer to the "flow" that Mihaly Csikszenmihalyi articulates (as discussed in Chapter 1), where what occurs is a fused, playful communication between two participants. In this participation, the user and the nanoscale communicate, and this communication is not one that necessarily ends with the creation of an image, as such habits develop capacities of response in the user.

This focus on alterations of rhetorical practices aligns with an understanding of writing as embodied response, or, as Marilyn M. Cooper has recently argued, for "embodied interaction with other beings and our environments" (18). Drawing on complexity theory and theories of technology, Cooper suggests that writing and technology are "cognate practices"—they "play the same role in our lives" in that they help us extend our understandings beyond our individual selves (18). Cooper argues that understanding writing and technology as cognates leads to a focus on practice and interaction, leading also to the development of rhetorical strategies that are useful with a given technology and situation (26–29). Cooper's articulation on practice and interaction aligns with the composition processes of STM images and the emphasis on image as experience that emerges from this study. Following Cooper, how might we understand imaging and technology as cognate practices (like writing and technology), especially in light of Barad's articulation of the deeply entangled nature of science, knowledge, and ourselves? If we do not consider the frame of analysis as part of our analysis, we risk missing how images like STM images—and other rhetorical communication—

are persuasive in their constitution of not only a "message," but also in their constitution of science, of rhetoric, and of ourselves.

Another way to highlight the rhetorical aspects of digital informational images like those the STM produces is to situate STM images in their context as new communication technologies. Some examples of how rhetorical strategies have shifted at the same time that new communication technologies have started to be used suggest that as technologies have expanded either the range of communication or the range of resources available to communicate, rhetorical practices have at times also expanded or shifted. One productive analytic frame relies on the rhetorical canons of invention, arrangement, style, memory and delivery. As Kathleen Welch observes in *Electric Rhetoric*, canons "have recurred in different forms and with different emphases in varying historical eras" (144). Building on Welch's observation, Brooke argues that canons should be seen as practices that also "always occur within a particular technological context" (xiii). Highlighting the ways in which practices associated with the canons may shift within the context of the introduction of new technologies, then, reveals some possible places for further analysis of what become sites of rhetorical practice in digital informational images, such as those the STM produces.

One example of shifts in rhetorical practices associated with the canons occurs around the emergence of alphabetic writing and rhetoric in fifth and fourth century B.C.E. in Athens, in using writing to expand oratory instruction. As Rosalind Thomas documents in *Orality and Literacy in Ancient Greece*, when alphabetic writing emerged in Athens in the eighth century B.C.E., Athenians did not widely adopt it. At first, the Athenians used writing as they used pictures, as memory aids or funerary markers, but by around the fifth century, they started using writing for other purposes. The sophists, a group of itinerant teachers of oratory, adopted writing as an aid in teaching the rich, young men of Athens how to give speeches so that they could defend themselves and their property in court. The sophists, often called the first professional teachers, offered "a kind of selective secondary education" (Kerferd 17). The sophists used new pedagogical methods to introduce expanded, advanced oratorical training to those Athenians who could pay them. While Thomas notes that in the fifth century, written documents had no intrinsic value, and were considered disposable memory aids, somewhere in the fourth century, writing gained value for different functions, such as the use of writing in developing arguments (92). In the fourth century B.C.E., an orator's use of the technology of writing became an additional, explicit goal in some sophistic pedagogy. For example, in a passage detailing characteristics of a good orator or statesman, the sophist Isocrates observes that formal training "cannot fully fashion men who are without natural aptitude into good debaters or writers" (15).

The introduction of writing into the sophists' teachings changed not only their methods, but also the rhetorical strategies they taught, given the new abilities that writing allowed those making arguments. For example, alphabetic writing afforded both writers and readers distance from each other so that both writer and audience are not necessarily face-to-face at the time of an argument's delivery. Writing also increased the time spent in composing by allowing revision on a much more intense scale, encouraged precise wording, and led to the production of writing to be regarded as a cultural object. The affordances of alphabetic writing also led to shifts in all five canons of rhetoric. Alphabetic writing allowed an increase in time that could be spent on invention of one text, instead of only on practicing heuristics or commonplaces; allowed more specific attention to arrangement and style of words and parts of a speech, as well as the overall speech; altered the canon of memory by creating an alternate store for memory than the biological individual; and altered delivery by focusing on the delivery in the text, as opposed to extra-linguistic attributes such as tone, gesture and expression.[104]

The technology of alphabetic writing afforded those teaching oratory a few important techniques. The use of expanded techniques to teach oratory also changed the value and status of the technology of writing, expanding the medium of argument to writing as well as oratory, and also altered what rhetorical training included. Changes such as the expansion of what writing was for suggest that new emphases occurred at least alongside the introduction of the alphabetic writing technology. As communication strategies can now include images, including STM images, for example, the sophists' adoption of alphabetic writing and the shift in rhetorical training practices suggest the possibility of a parallel occurrence.

For example, like the technology of writing, as my study of the STM has shown, the technology of the digital informational image increases the time available for invention, as image creators interact with the image not only through the STM, but also through imaging programs like *Photoshop*. Indeed, the variety of possibilities for imaging also increase, as the image-creator's intent is not to present a view of an object, but to present data in an arrangement that conveys the image-creator's points about the significance of that data in a credible, persuasive way. Thus, too, arrangement is emphasized: Arrangement techniques can apply to the data overall, but can also apply to parts of the image, for image-makers may focus on arrangement pixel by pixel. As *Haptic Visions* has elaborated, the affordance of extended time to create the STM image and interact with the data leads to techniques of arrangement that do not merely focus on representation, but also emphasize the experience of the viewer as he or she follows the data. To be persuasive, arrangement also includes attention to readers' responses and, in the case of image-creators, includes their responses to the data. Delivery, too, becomes extended as the experience of the image becomes paramount: Delivery becomes not only delivery of an argument, but also delivery of an experience. Delivery and invention become fused, as readers invent as they engage in the reading process, especially if the images are computer-based.

While comparisons between adoptions of new technologies do not draw definitive parameters for a current-day shift in the canon (certainly not my intention),[105] comparing the adoptions of technologies suggests possibilities for considering the rhetorical work images accomplish in terms of the larger context of how digital images like STM images may also affect rhetorical knowledge and practice. Comparisons also suggest possibilities for considering techniques used to create communication in light of their rhetorical and knowledge-creating capacities: How are techniques of creating images also changing the value and status of visualization technologies as communication technologies? How are image-creation technologies changing the value and status of images as forms of communication? As we pursue questions like these about the impact of the process of creating communication, it is important to consider not only what technologies afford, but also the context in which imaging practices occur. As Brooke reminds us, for example, the history of hypertext shows that hypertextuality has permeated media in ways different than those early in the history of hypertext expected (xii).

The analysis in this book also suggests that the heightened interaction necessary to make nanoscale phenomena visible, and atoms seem tangible, also changes us. The understanding of the nanoscale that STM images present emphasizes a process in which haptic human interaction with atomic phenomena is overt and necessary in order to learn about the nanoscale—we understand as we touch, as we immerse ourselves within the data, within the experience of the microscope. Indeed, we immerse ourselves within haptic vision practices that we take with us as we communicate about not only what and how we have seen, but also how we have changed. We can't help but touch and be touched in the process of discovery and communication—a far cry from the figure of the distant, objective observer of old. In order to learn and communicate, we must respond and respond again. We must reach out with our hands as well as our eyes.

Notes

Introduction

1. The two versions of Eigler and Schweizer's final "IBM" image found, in Figure 1 and Figure 2, have appeared on *Good Morning America*, CNN, the NBC news, in the AP wire (Lovejoy), and in *Time* (Stengel 40–42), as well as countless online citations. Textual accounts include popular histories (see Regis 11, 232) and sci-fi nanothrillers (see Crichton 133). Martina Merz presents an interesting philosophical analysis of how the "IBM" images communicate differently to scientists and non-scientists.

2. Many trace future visions to K. Eric Drexler's *Engines of Creation* and Bill Joy's "Why the Future Doesn't Need Us." These texts include an assortment of future visions that continue to appear in nanotechnology discourse, including in policy arguments, for example (see W. P. McCray on utopian visions in U.S. governmental policy statements and Valerie Hanson on appearances in a U.S. House of Representatives hearing).

3. See McCray for examples of arguments; also see Hanson. Funding sources include, for example, U.S. government multi-agency National Nanotechnology Initiative, whose budget in 2014 is $1.7 billion (United States, National Nanotechnology Initiative, "NNI Budget").

4. This is a general definition: The National Nanotechnology Initiative defines nanotechnology as

> the understanding and control of matter at dimensions between approximately 1 and 100 nanometers, where unique phenomena enable novel applications. Encompassing nanoscale science, engineering, and technology, nanotechnology involves imaging, measuring, modeling, and manipulating matter at this length scale. (United States, National Nanotechnology Initiative, "What It Is")

5. For some examples, see McCray; also see Hanson.

6. Social scientists have explored some aspects of nanotechnology discourses. Specifically, see: Sarah Kaplan and Joanna Radin and Cynthia Selin for interesting analyses of how arguments over the field's definition have shaped nanotechnology; Brenton Faber on the rhetoric of popular media accounts of nanotechnology; David Berube on the rhetoric of risk communication in nanotechnology; and Hanson on the rhetorical use of future visions in arguments about funding societal and ethical implications research.

7. Arguments for images as important components of scientific practice have become common in science studies and related fields since Martin Rudwick's 1976 "The Emergence of a Visual Language for Geological Science." Lynch and Steve Woolgar's edited collection contains several influential, now-classic arguments for the study of images; also see Luc Pauwels's edited collection for relevant articles; and see James Elkins, *The Domain of Images*, for a survey of scholarship.

8. Also see Sabine Brauckmann, who follows Rudwick's lead and demonstrates how images helped establish developmental biology.

9. See Timothy Lenoir, "Shaping Biomedicine," for an argument about the configuration of biology as an information science (27). A number of scholars have commented on the digitization of the humanities. See, for example, N. Katherine Hayles, *How We Think*; Jeff Rice; Anne Burdick et. al; Johanna Drucker, *Speclab* and "Reading Interface," one of a series of essays on the topic published in the January 2013 edition of *PMLA: Publications of the Modern Language Association of America*. Also, see Paul Wouters et al. for a collection of essays documenting how knowledge practices are changing in the humanities and social sciences. Also see Nigel Parton, for example, for discussion of the impact of new information and communication technologies on social work.

10. "New media," or digital media studies, tend to focus on film, new media art, or literature, as opposed to science. The tables of contents of popular readers like *The New Media Reader* (edited by Noah Wardrip-Fruin and Nick Montfort), for example, reflects this tendency. Also see Hayles, *Writing Machines*; Anna Munster; and Jacques Khalip and Robert Mitchell. In this book, I use "digital media" instead of "new media," unless I refer to the specific work of a scholar who uses the term, as "digital media" is the more specific of the two terms.

11. Informational images, or data-rich images, have been studied from a number of perspectives: for example, as "non-mimetic representations" (Daston and Galison 349); "non-isomorphic images" (Ihde, 37); or as "non-art" or "informational" images (Elkins, *Domain*): each of these appellations suggests some of the important qualities of these images, whether they are charts, graphs, or other graphic presentations of data. Other useful sources about informational images include the works of Edward Tufte, including *Envisioning Information*.

Chapter 1

12. "B." Personal interview. 6 July 2005. This interview (and others cited) were conducted as a series of anonymous interviews as part of research approved by the Philadelphia University Institutional Review Board.

13. This account is partial because it is beyond the scope of this chapter to fully develop a detailed history of this transition. For historical accounts of the development of the STM, see Cyrus C. M. Mody, *Instrumental Community*; Mody and Lynch; and Galina Granek and Giora Hon; also see Gerd Binnig and Heinrich Rohrer, "From Birth," for an account of the STM by its inventors. For a history of concepts of and discoveries about the atom for a popular audience, see Hans Christian Von Baeyer.

14. See Ollie Oviedo, Joyce Walker, and Byron Hawk's edited collection, especially Walker and Hawk's introduction, for examples of a focus on technologies and tools of computer-mediated communication that accounts for the rhetorical aspects of instruments within knowledge-producing systems.

15. Joseph Dumit, Morana Alac, Kelly Joyce, Aug Nishizaka, Hans Rystedt et al., and Monika Büscher and Gloria Jensen published recent examples of laboratory studies featuring visualization technologies. Classic laboratory studies include Bruno Latour and Steve Woolgar's *Laboratory Life*, Lynch's *Art and Artifact*, and Klaus Amann and Karin Knorr Cetina's "The Fixation of (Visual) Evidence," for example.

16. Many science studies scholars argue that material and social practices significantly affect the shaping of scientific knowledge. For example, see Galison, *Image and Logic*; Latour, *Science in Action*;

Lenoir, *Instituting Science*; Nelly Oudshoorn; and Steven Shapin and Simon Schaffer.

17. Nicolas Rasmussen provides a good example of this process of habituation, discussing how electron microscopists become habituated to using the instrument, although he does not explore the effects of this habituation on rhetorical practices (227–32).

18. In the early days of the STM, the STM image was created by tracing a pen on paper. The researcher then compiled the tracings to form an image (Park and Barrett 64). However, the method of creating a data matrix is similar either with or without the help of computer graphics.

19. For further examples of this mediation, see Dumit for an account of PET scans and Alac for an account of fMRI.

20. This apparent simplicity also sparks debates about images' as scientific, medical, or legal evidence. See Tal Golan for the use of X-rays as legal evidence and Dumit (Chapter 4) for the use of PET scans in courtrooms.

21. This confusion, of course, also extends to non-experts. See Adina Roskies for lay readers' confusion about fMRI.

22. Galison claims the fact that the data-image has become accepted indicates a shift in traditions and in science: "Increasingly, especially for the younger experimenters of the 1980s and 1990s, these [image vs. logic] are distinctions without meaning; the fragile trading zone of the 1970s between the two traditions has become the site of a new generation of experimentation" (*Image and Logic* 810).

23. See Kelly Joyce for an account of how the use of images to present data using MRI was developed (33–34); see Mody and Lynch for an account of some of the corporate drivers behind the development of new instruments in semiconductor surface science, including computer capacity (440-443). Also, visualization technologies like MRI began to be used in settings outside the laboratory, such as in hospitals in the 1970s (K. Joyce 9).

24. Mody traces the more specific development and establishment of the STM in *Instrumental Communities*. He provides a fascinating account of the development of the instrument within disciplines and other communities.

25. Studies of other data presentations do exist, such as narrative visualizations in infographics appearing in newspapers (Hullman and Diakopolous), or more historical data images such as statistical

atlases (Kostelnick, "Melting-Pot") or Florence Nightingale's "Rose Diagrams" (Brasseur).

26. Some recent science studies ethnographies consider the embodiment of meaning-making. For example, Alac analyzes the MRI and its productions in the process of scientific inquiry, and documents how researchers make meaning from MRI brain scans through gestural interaction. Natasha Myers argues for role of gesture in researchers' understandings of protein folding. Matthew Kirschenbaum's inclusion of the material components of an instrument—the computer—in his analysis of electronic texts provides a rich model from textual studies. Rhetoricians have also turned to include embodiment in studying digital composition. Kristin Arola and Anne Frances Wysocki's recent collection, for example, shows a wide variety of careful considerations of the effects of digital media on our physical, lived experience.

27. My close reading also extends the study of interactive practices in science studies, as, for the most part, ethnographers of scientific practice focus on interactions only in the processes of making meaning from the productions of an instrument. For example, see: Alac for analysis of how fMRI brain scans form sites of interaction (17–18); Nishizaka for a study of how the view of a fetus is constructed in a medical setting through ultrasound and the interactions between medical practitioner and woman receiving the ultrasound; Amann and Knorr Cetina for analysis of seeing visual inscriptions as interactive when embedded in talk (92).

28. See Dominik Leiner and Olivier Quiring for a brief overview of social scientists' disagreements (128); see Lev Manovich for a dismissal of the term as a redundant description of structure of the computer in the widely-cited *Language of New Media* (55–56).

29. See Mody, *Instrumental Community* (Chapter 2) for a history of the topografiner. Mody argues that the STM was successful where the topografiner was not because the inventors of the STM used their organizational and disciplinary networks more effectively. Binnig and Rohrer acknowledge the close connection of the instrument in their Nobel prize speech: "Had they [Young, Ward, and Scire], even if only in their minds, combined vacuum tunneling with scanning, and estimated that resolution they would probably have ended up with the new concept, Scanning Tunneling Microscopy" (Binnig and Rohrer, "From Birth" 392). Binnig and Rohrer also state that they only first heard of the topografiner two years after having the idea for the STM,

and "shortly before getting [their] first images" ("From Birth" 392), illustrating the development of multiple visualization technologies at the time.

30. It is, of course, hard to tell if only one atom exists, because the STM is the first microscope to consistently visualize atoms; therefore, researchers often guess and find out if they have only one atom by looking at the image the STM creates—yet another interaction that structures responses between researcher and atomic phenomena.

31. "B." Personal interview. 6 July 2005. Also, physicist D. P. Woodruff observes that adsorbed atoms and molecules move more rapidly than the movement of the STM tip, and so, visualizing them is widely seen as a problem (77). The "scan rate" has been identified in the 2006 Board on Chemical Sciences and Technology report as the main limitation of the STM (121).

32. Also see Mody, *Instrumental Community*; and Hennig, "Changes" for more on the history of STM image development.

33. "C." Personal interview. 2 Aug. 2005.

34. The report defines chemical imaging as "the spatial (and temporal) identification and characterization of the molecular chemical composition, structure, and dynamics of any given sample" (Board on Chemical Sciences and Technology 14).

35. "E." Personal interview. 22 Aug. 2005.

36. As one researcher I interviewed explains, "You can filter images and make them look cleaner but, in fact, we always try and take our data and prepare our instruments that are stable enough that the raw data is what we present in a paper, and if we ever do something different, then we say that specifically." ("B." Personal interview. 6 July 2005).

37. Also see Amann and Knorr Cetina; Dumit; Alac; and Kelly Joyce, for further examples.

38. "Drift" is thought to be due to the movements of atoms and molecules that are faster than the scanning speed (Board on Chemical Sciences and Technology 118).

39. "G." Personal interview. 22 June 2006.

40. "G." Personal interview. 22 June 2006.

41. "C." Personal interview. 2 Aug. 2005.

42. Scientists may create images for contests run by companies and other institutions, such as the SPM instrument company Veeco.

See, for example, "Veeco SPM Calendar Image Competition," or the California Academy of Sciences's Photo Competition.

43. "G." Personal interview. 22 June 2006.

44. "B." Personal interview. 6 July 2005.

45. "I have a file. I dump many images in and then I leave it for a while, for one or two days, and when I have a peaceful time I look back again and which one I like and I choose those images" ("F." Personal interview. 23 June 2006).

46. Dumit provides an example of how color choices can affect how PET scans are interpreted, examining a series of differently colored images produced from the same data that lead to different interpretations of the data (91).

47. "G." Personal interview. 22 June 2006.

48. "B." Personal interview. 6 July 2005.

49. For discussion of the impact of this theory on the basis of cybernetics, see Hayles, *How We Became Posthuman*. For impacts on molecular biology, see Richard Doyle. It also connects to the quantum mechanics view of the universe, as Stefan Helmreich explains: "[a]ccording to quantum mechanics, the only thing that truly exists in the universe is pattern; the substance that supports these patterns is, in some fundamental sense, inessential" (76).

Chapter 2

50. This view of the role of metaphor has been argued by many in a wide range of fields, such as rhetoric, cognitive science, and science studies. For example, see George Lakoff and Mark Johnson; Max Black; Theodore Brown; N. Katherine Hayles, *How We Became Posthuman*; Evelyn Fox Keller, *Refiguring Life*; and Keller, *Making Sense of Life*. Also see Chapter 4 for more on metaphors.

51. Hal Foster differentiates between "vision" and "visuality" by describing the former as sight as a physical operation that is also social and historical, and the latter as sight as a social fact that also involves the body and psyche (ix). While others have followed this differentiation to emphasize the socially constructed nature of some of our seeing practices (see Laura U. Marks, for example), here I use "vision" instead of "visuality" to push against the common linkage between one kind of vision practice (i.e., perspectivalism) and the physical operation of sight, a linkage I discuss further below. Above all else, I want

to emphasize that the practices I describe here are physical, social, and historical.

52. Here I follow Gilles Deleuze and Felix Guattari's distinction: "'Haptic is a better word than 'tactile' since it does not establish an opposition between two sense organs but rather invites the assumption that the eye itself may fulfill this nonoptical function" (492).

53. See, for example, George Landow; Michael Joyce; and Jay David Bolter (35–36).

54. For accounts in art history, see Alois Riegl and his followers, including Wilhelm Worringer. In film studies, scholars who specifically mention the haptic in recent work, often following Riegl, include: Laura U. Marks, Jennifer Fisher, and Antonia Lant. Digital or new media studies has also taken up the haptic: See Mark B. N. Hansen, *New Philosophy for New Media.*

55. In one recent example, Jeanne Fahnestock argues for the inclusion of visual as well as textual documents in the study of scientific discourse, arguing that "scientific arguers often resort to visual persuasion" by including visuals as part of their argument, and suggests that rhetorical elements are expressed consistently in both visual and textual discourse (*Rhetorical Figures* xi).

56. For example, see Beverly Sauer.

57. Science studies scholars have explored the position of the observer in different contexts. For example, Michael Lynch describes two different spatial orders that he argues constitute a physical space, or "topical contexture" in which scientists operate in the laboratory, and that employ differing dynamics of seeing ("Laboratory Space" 56, 73-74). Nicolas Rasmussen also considers practices in a brief discussion of electron microscopists' frequent shifts towards Don Ihde's category of embodiment (where an instrument user sees *through* the instrument, as opposed to *seeing* the instrument) in the microscopists' process of learning how operate the electron microscope (227–32). Peter Galison and Lorraine Daston and Galison describe some of the possible shifts in the role of the scientist-observer given different interactions with images. Also, Keller discusses the conventional microscope user's coordinated hand-eye practices, and argues that these practices have influenced the study of life by spurring such interventionist methods as introducing fluorescent tracers via viral vectors into genes, altering the objects of study as well as the methods of interaction scientists use ("Biological"118-21). The attention to the observer in the work of

these scholars suggests the complexity of the interactions present in visualization technologies, and some of the importance of studying how users and visualization technologies are affected by and interact with vision practices.

58. My intention here is not to counter perspectival vision or to show it is a "mistake." Indeed, practices falling under this title are extremely useful and important. Instead, like Lynch, I aim to note that the stationary observer is a position within a cultural—and not entirely biological—complex (Lynch, "Lab" 58). As Lynch also notes, the stationary observer learns to assume a fixed position that becomes an embodied relation: "The passive subject of the contemplative approach becomes an embodied relation and thus a subject that locally validates the opticist picture" ("Lab" 61).

59. Art historical studies about culturally and historically specific vision practices (besides Riegl and Crary) include Michael Baxandall's study of fifteenth-century Italian vision practices and Svetlana Alpers's description of the sixteenth-century Dutch pictorial style.

60. In light of this assumption, it is particularly interesting that the STM was not originally conceived as a microscope, but as a spectroscope. As its inventors Binnig and Rohrer admit, they had no experience in microscopy or surface science, although they did have backgrounds in vacuum tunneling and Ångstrom-level work; they comment: "This probably gave us the courage and light-heartedness to start something which should 'not have worked in principle' as we were so often told" ("From Birth" 389).

61. Hooke's preface to his *Micrographia* presents a particularly clear case of this, as he describes building up a picture over a series of observations of the same object. Also see Ford (Chapter 8) for an intriguing photograph of what Hooke most likely saw through the microscope: Comparison of the engravings shows that Hooke's view and Hooke's engravings present different images.

62. Lev Manovich discusses how radar becomes a fundamentally new technology through the screen it uses to show information:

> What is new about such a screen is that its image can change in real time, reflecting changes in the referent, whether the position of an object in space (radar), any alteration in visible reality (live video) or changing data in the computer's memory (computer screen). The image can be continually updated in real time. (99)

63. Here I do not argue that the other loses otherness (alterity), or that we are all the same, but that everything in the equation changes, including the observer's subjectivity.

Chapter 3

64. Researchers could easily communicate this to the illustrator if the scientists did not generate the image themselves, as scientific illustrators work closely with the scientists whose work they portray. See, for example, Frank Ippolito for a scientific illustrator's response to Ottino.

65. Cyrus C. M. Mody observes that scientists realized that even misinterpreted images may generate interest if the subject was one that interested a scientific community (*Instrumental Community* 114–15).

66. See Alberto Cambrosio, Daniel Jacobi, and Peter Keating for an example of how Ehrlich's diagrams took on a role in scientific practice (666–67). Also see Michael Ruse.

67. While STM images also participate within other arguments such as journal articles, it is beyond the scope of this chapter to conduct analysis of both visual and verbal elements. Chapter 4 assesses tropes running through both visual and verbal discourses as a start to this area of analysis, although further work would enrich this understanding of nanotechnology discourses.

68. "B." Personal interview. 6 July 2005.

69. In this chapter, I focus on images made for publication, what Klaus Amann and Karin Knorr Cetina call "evidence": visuals designed to be included in scientific papers, in their categorization of three modes of practice in laboratory work that involve visuals (88).

70. In fact, one issue that emerges from the use of images to communicate data is their capacity to express so much data. Greg Myers explores the communicative properties of scientific images by charting how much "gratuitous detail" they contain; such detail can obscure the main message of the image (234).

71. Also see Steve Woolgar for discussion of how data and visuals are made sense of by referring to previous experiences or documents.

72. See Stephen Bernhardt as well as Karen Schriver, for example.

73. For examples of comparing image elements to language, see Charles Kostelnick and David Roberts; Hanno Ehses; Jeanne Fahnestock, *Rhetorical Figures*; Keith Kenney; and Caroline van Eck for ex-

amples of how traditional (verbal) rhetorical concepts can be seen in visual contexts. For examples of identifying a semiotic grammar of visual images, see Gunther Kress and Theo van Leeuwen; Claire Harrison; and Maureen Daly Groggin.

74. See J. Anthony Blair, for example. Also see Gross for a summary of this debate ("Toward A Theory").

75. As Elkins eloquently argues, although Alberti advocated his method as the only method of perspective, no one unified system existed. Many artists and writers had their own variety of perspective. It is only in modern times, Elkins argues, that "perspective" has flattened to one model, based on Alberti (Elkins, *The Poetics of Perspective* xi, 217-261). Here I refer to the more modern "scopic regime" because my point about these images is how they depart from this dominant regime.

76. Martin Jay, for example, uses the term in *Downcast Eyes*. Jay takes this term from film critic Christian Metz.

77. What is also fascinating about this processed image is that it is actually the second "image," the first one being an actual object that was created by gluing together the original line recordings of the STM (Binnig and Rohrer, "From Birth" 400).

78. See, for example, Matthew Chalmers, Douglas Mulhall (40), and Chad Mirkin. Also, Chapter 4 explores this point further, with more specific analysis of the common scientific trope of small or invisible worlds.

79. "C." Personal interview. 2 Aug. 2005.

80. "C." Personal interview. 2 Aug. 2005.

81. G." Personal interview. 22 June 2006; "B." Personal interview. 6 July 2005. See Chapter 1 for further detail on these color choices. Also, the use of false color functions in other fields: see Lynch and Samuel Y. Edgerton's reports on the lack of organized use of color in astronomy, for example ("Aesthetics"); also see Joseph Dumit for discussion on color in PET images.

82. It is significant that this intensified attention to color appears in images not designed for journal article texts, indicating, perhaps, a rhetorical strategy for providing a clearer guide through informational images to readers less familiar with reading STM images. However, the combination of linear perspective and image-interface conventions also occurs in black-and-white images in journal articles, too—two of the more widely-circulated images, black-and-white corral images

and the "IBM" images (Figure 1), also show the attention to read the surface via contours and texture of the data, calling forth the fingertips and the eyes.

83. As Deleuze observes, each or a group of the bits composing a digital image can in themselves create an image, allowing for unlimited production of other images from the bits (*Cinema 2* 265).

Chapter 4

84. This narrative occurs in various summaries of the origins of nanotechnology. For example, it appears in the National Nanotechnology Initiative's account of nanotechnology (United States, NNI National Nanotechnology Initiative, "What It Is") and in the National Science and Technology Council's brochure intended for the public, "Nanotechnology: Reshaping the World Atom by Atom." It also appears in general science journals geared towards scientists, such as in *Science*'s special issue on nanotechnology, *Science* 254 (November 29, 1991), and is mentioned in individual articles such as James Gimzewski and Christian Joachim's, for example. The narrative's accuracy in relating the development of nanotechnology has recently been questioned. See, for example, Chris Toumey's examination of the citations of Feynman's speech and recollections of some of the scientists involved in major nanotechnology discoveries. Colin Milburn has also pointed out many uncited references to science fiction in Feyman's speeches (*Nanovision* 37, 47–49). For another history of the development of nanotechnology that accounts for the role of probe microscopists, see Cyrus C. M. Mody, *Instrumental Community* (Chapter 6). Mody also discusses STM co-inventor Heinrich Rohrer's assessment that in addition to new instruments, new science governance entities helped shape nanotechnology (*Instrumental* 189).

85. See Toumey for an account of the citation history of the speech.

86. The only other possible origin text discussed by nanotechnology proponents seems to be Einstein's doctoral dissertation, where he calculates the diameter of a sugar molecule as one nanometer.

87. For example, see Jose Julian Lopez for a recent discussion of metaphors in general as well as in genomics ("Notes"). See also Brigitte Nerlich, Richard Elliott, and Brendon Larson for discussions of bio-

logical metaphors in relation to ethics and journalism in science. See Andrew Ortony for a collection of studies on metaphor.

88. Luc Pauwels's recent edited collection brings together some important essays on this topic, and Michael Lynch and Steve Woolgar present another major collection on this topic. Also see James Elkins's *Domain of Images* (Chapter 1) for a summary of some of the important research on scientific images.

89. See Andreas Lösch as well as Alfred Nordmann, for example, for other useful analyses of images of nanotechnology that do not explicitly discuss metaphor. Nordmann's analysis of a few key nanotechnology images presents parallels with the ability of metaphors to shift given cultural changes. Although Nordmann does not discuss images in terms of metaphor, he explains that an image of nanotechnology accrues additional associations as the field of nanotechnology develops.

90. Michelle Sidler makes this point about molecular biology, arguing that all sciences that rely more on information technology to convey information about objects of study also rely more on metaphor (61). Lopez provides an example of the explanatory power of metaphor in nanotechnology in his study of how the metaphor of nanotechnology practitioners as master builders functions to organize multiple types of narratives in nanotechnology ("Nanotechnology").

91. For more on the contested issue of defining nanotechnology, see: Cynthia Selin; Ashley Shew; and Lopez, "Nanotechnology" (1270–73).

92. Although Eigler and Schweizer published a xenon atom line as well as the "IBM" images in the same *Nature* article, the line of atoms was published on the last page, and was not cited as "the first" evidence of manipulation (526). In fact, neither series of images was mentioned as "the first." The line image has not circulated as widely as the "IBM" image, and is republished as the first instance of manipulation in only a handful of instances (see Joseph A. Stroscio and Eigler, Fig 6(a), 1324, for one example).

93. Much has been written about how writers from the ancient Greek atomists to the present have used the concept of the "book of nature" to describe phenomena or their mastery of them. For an analysis of ancient atomists' use of this metaphor, see Fernand Hallyn. For other analyses of this metaphor, see Jacques Derrida (*Of Grammatology* 15) and Richard Doyle (Chapter 3), for example.

94. Susan Stewart recounts that the practice of micrographia may have existed well before print, but certainly became a fad around the invention of the printing press (38). It is interesting that Feynman should call for new instantiations of tiny writing at the beginning of what we could call "the computer age," in light of Stewart's comment that such tiny writing in book form became a fad right at the introduction of the printing press, partly because such tiny books emphasized the human craft necessary to produce them in limited numbers, in contradistinction to the ability of the printing press to produce many copies with less human attentiveness (Stewart 38).

95. In fact, Eigler had not paid attention to Feynman's speech until after he and Schweizer produced the "IBM" images (Appenzeller 1300).

96. The most recent bout between Drexler and Smalley occurs in the December 2003 issue of *Chemical and Engineering News* (Baum 37-42), and was recounted in a *New York Times* article (Chang D3). For an early critical review of nanotechnology, particularly of Drexler's ideas, see Garfinkel, as it contains written exchanges with Drexler. Ed Regis provides a popular historical account. Also see Sarah Kaplan and Joanna Radin for a thorough and fascinating analysis of the entire debate and its functions within the complex negotiation of the authority to define nanotechnology.

97. Some examples include E. P. Stoll (68); J. G. Kushmerick et al.; Binnig and Rohrer, "Scanning Tunneling Microscopy" (361); and C. T. Salling and M. G. Lagally (503).

98. This quote has also been reprinted in a *Popular Science* article about the scanning tunneling microscope and the Nobel Prize to describe Binnig and Rohrer's feelings of accomplishment when they imaged the "7x7 Silicon," and is identical to Binnig and Rohrer's use of the metaphor in their Nobel speech (Fisher, "Seeing Atoms" 106). Also see Milburn for further examples and an argument about the links between the claims of nanoscience to see a small world and to be able to see and control the future (*Nanovision* 62, 83).

99. See Brian Ford (Chapter 8) for an intriguing photograph of what Hooke most likely saw through the microscope: Comparison with the engravings shows that Hooke's view and Hooke's engravings are dissimilar in many ways.

100. See Milburn for discussion of the "corral" connection to Manifest Destiny and western expansion (*Nanovision* 67).

101. Milburn reads this corral construction as "a scientific mapping practice, an effort to contain novel territory within a representational topography . . . that transforms its various physical properties into property as such" (*Nanovision* 65–66). While my reading does not directly contradict this view, my focus on trope emphasizes the importance of the built structure as part of the "subvisible worlds," especially in comparison with the conventional microscope trope.

102. See Milburn, "Atoms and Avatars," for an explanation of the inclusion of nanotechnology in virtual worlds; see Lorraine Daston and Peter Galison for a discussion of these "images-as-tools" as a successor to visual atlases as collections of knowledge (383–412).

Conclusion

103. This point resonates with other observations about contemporary science, such as Timothy Lenoir's argument that due to the fact that biologists were able to gather much more data from new technologies, such as "cloning, restriction enzymes, protein sequencing, and gene product amplification," the field of biology became an "information science" in the mid-1960s, changing the organization of the field (Lenoir, "Shaping" 27).

104. For more information on these changes, see (for example): Thomas; Isocrates; G. E. Kerferd; LaRue Van Hook; Susan Jarratt; Nikos Poulakos; and H. I. Marrou.

105. I am not arguing for a revision of the canon. See Collin Gifford Brooke for an interesting take on re-envisioning the canon in light of new media.

Works Cited

Alac, Morana. *Handling Digital Brains: A Laboratory Study of Multimodal Semiotic Interaction in the Age of Computers.* Cambridge: MIT P, 2011. Print.

Alpers, Svetlana. *The Art of Describing.* Chicago: U of Chicago P, 1983. Print.

Amann, Klaus, and Karin Knorr Cetina. "The Fixation of (Visual) Evidence." Lynch and Woolgar 85–122.

Appenzeller, Tim. "The Man Who Dared to Think Small." *Science* 254 (1991): 1300. Print.

Arola, Kristin L., and Anne Frances Wysocki, eds. *Composing(media) = Composing (embodiment).* Boulder: UP of Colorado, 2012. Print.

Bacon, Francis. *The New Organon: Or True Directions Concerning the Interpretation of Nature.* Trans. J. Spedding, R. L. Ellis, and D. D. Heath. Boston: Taggard and Thompson, 1863. Web. 29 July 2013.

Bai, Chunli. *Scanning Tunneling Microscopy and its Applications.* 2nd ed. Berlin: Springer-Verlag, 2000. Print.

Barad, Karen. *Meeting the Universe Halfway: Quantum Physics and the Entanglement of Matter and Meaning.* Durham: Duke UP, 2007. Print.

Bastide, Françoise. "The Iconography of Scientific Texts: Principles of Analysis." Lynch and Woolgar 187–230.

Bateson, Gregory. *Steps to an Ecology of Mind.* New York: Ballantine, 1972. Print.

Baum, Rudy. "Nanotechnology: Drexler and Smalley Make the Case For and Against 'Molecular Assemblers.'" *Chemical and Engineering News* 81.48 (2003): 37–42. Print.

Baxandall, Michael. *Painting and Experience in Fifteenth-Century Italy: A Primer in the Social History of Pictorial Style.* Oxford: Oxford UP, 1972. Print.

Bear, Greg. *Eon.* New York: Tor, 1985. Print.

—. *Eternity.* New York: Warner, 1988. Print.

Bechtold, Adrian, Peter Hadley, Takeshi Nakanishi, and Cees Dekker. "Logic Circuits with Carbon Nanotube Transistors." *Science* 294 (2001): 1317–20. Print.

Becker, R. S., J. A. Golovchenko, and B. S. Swartzentruber. "Atomic-scale Surface Modifications Using a Tunnelling Microscope." *Nature* 325 (1987): 419–21. Print.

Benjamin, Walter. "The Work of Art in the Age of Mechanical Reproduction." *Illuminations*. Ed. Hannah Arendt. Trans. Harry Zohn. New York: Schocken, 1968. Print.

Berger, John. *Ways of Seeing*. London: BBC and Penguin, 1972. Print.

Bernhardt, Stephen. "Seeing the Text." *College Composition and Communication* 37.1 (1986): 66–78. Print.

Berube, David. "The Rhetoric of Nanotechnology." *Discovering the Nanoscale*. Ed. Davis Baird, Alfred Nordmann, and Joachim Schummer. Amsterdam: IOS Press, 2004. 173–192. Print.

Binnig, Gerd, and Heinrich Rohrer."The Scanning Tunneling Microscope." *Scientific American* Aug. 1985: 50–56. Print.

—. "Scanning Tunneling Microscopy." *IBM Journal of Research and Development* 30.4 (1986): 355–69. Print.

—. "Scanning Tunneling Microscopy—from Birth to Adolescence." *Nobel Lectures, Physics 1981–1990*. Ed. Tore Frängsmyr and Gösta Ekspång. Singapore: World Scientific Publishing, 1993. 389–409. Web. 29 July 2013.

Binnig, Gerd, Heinrich Rohrer, Christoph Gerber, and Edmund Wiebel. "Tunneling Through a Controllable Vacuum Gap." *Applied Physics Letters* 40.2 (1982): 178–80. Print.

Black, Max. *Models and Metaphors: Studies in Language and Philosophy*. Ithaca: Cornell UP, 1962. Print.

Blackmore, Susan. *The Meme Machine*. Oxford: Oxford UP, 1999. Print.

Blair, J. Anthony. "The Rhetoric of Visual Arguments." Hill and Helmers 41–61.

Blanchot, Maurice. *The Gaze of Orpheus and Other Essays on Literature*. Ed. P. Adams Sitney. Trans. Lydia Davis. Barrytown: Station Hill, 1981. Print.

Board on Chemical Sciences and Technology. *Visualizing Chemistry: The Progress and Promise of Advanced Chemical Imaging*. Washington, DC: National Academies Press, 2006. Web. 29 July 2013.

Bogost, Ian. *Persuasive Games: The Expressive Power of Videogames*. Cambridge: MIT P, 2010. Print.

Bolter, Jay David. *Writing Space: Computers, Hypertext, and the Remediation of Print*. 2nd ed. Mahwah: Lawrence Erlbaum, 2001. Print.

"A Boy and His Atom: The World's Smallest Movie." *IBM Research*. IBM, n.d. Web. 23 July 2013.

Brasseur, Lee. "Florence Nightingale's Visual Rhetoric in the Rose Diagrams." *Technical Communication Quarterly* 14.2 (2005): 161–82. Print.

Brauckmann, Sabine. "On Fate and Specification: Images and Models of Developmental Biology." *The Educated Eye: Visual Culture and Pedagogy in the Life Sciences*. Ed. Nancy Anderson and Michael R. Dietrich. Hanover: Dartmouth College P, 2012. 213–34. Print.

Brauman, John. "Room at the Bottom." *Science* 254 (1991): 1277. Print.

Brodbeck, Dominique, Riccarde Mazza, and Denis Lalanne. "Interaction Visualization: A Survey." *Human Machine Interaction*. Ed. Denis Lalanne and Jürg Kohlas. Berlin: Springer-Verlag, 2009. 27–46. Print.

Brooke, Collin Gifford. *Lingua Fracta: Toward a Rhetoric of New Media*. New York: Hampton Press, 2009. Print.

Brown, Theodore. *Making Truth: Metaphor in Science*. Urbana: U of Illinois P, 2003. Print.

Buchwald, Jed Z. "How the Ether Spawned the Microworld." *Biographies of Scientific Objects*. Ed. Lorraine Daston. Chicago: U of Chicago P, 2000. 203–25. Print.

Burdick, Anne, Johanna Drucker, Peter Lunenfeld, Todd Presner, and Jeffrey Schnapp. *Digital Humanities*. Cambridge: MIT P, 2012. Print.

Burke, Kenneth. *Language as Symbolic Action: Essays on Life, Literature and Method*. Berkeley: U of California P, 1966. Print.

Büscher, Monika, and Gloria Jensen. "Sound Sight: Seeing with Ultrasound." *Health Informatics Journal* 13.1 (2007): 23–36. Print.

Cambrosio, Alberto, Daniel Jacobi, and Peter Keating. "Ehrlich's 'Beautiful Pictures' and the Controversial Beginnings of Immunological Imagery." *Isis* 84 (1993): 662–99. Print.

Canetti, Elias. *Crowds and Power*. Trans. Carol Stewart. 1962. New York: Farrar, Straus, and Giroux, 1973. Print.

Carnegie, Teena A. M. "Interface as Exordium: The Rhetoric of Interactivity." *Computers and Composition* 26 (2009): 164–73. Print.

Chalmers, Matthew. "Second Law of Thermodynamics Broken." *New Scientist* 19 July 2002. Web. 29 July 2013.

Chang, Kenneth. "Yes, They Can! No, They Can't: Charges Fly in Nanobot Debate." *The New York Times* 9 Dec. 2003: D3. Print.

Cooper, Marilyn M. "Being Linked to the Matrix: Biology, Technology, and Writing." *Rhetorics and Technologies: New Directions in Writing and Communication*. Ed. Stuart Selber. Columbia: U of South Carolina P, 2010. 15–32. Print.

Crandall, B. C. *Nanotechnology: Molecular Speculations on Global Abundance*. Cambridge: MIT P, 1996. Print.

Crary, Jonathan. *Techniques of the Observer: On Vision and Modernity in the Nineteenth Century*. Cambridge: MIT P, 1990. Print.

Crichton, Michael. *Prey*. New York: Harper, 2002. Print.

Crommie, M. F., C. P. Lutz, and D. M. Eigler. "Confinement of Electrons to Quantum Corrals on a Metal Surface." *Science* 262 (1993): 218–20. Print.

Curtis, Scott. "Photography and Medical Observation." *The Educated Eye: Visual Culture and Pedagogy in the Life Sciences*. Ed. Nancy Anderson and Michael R. Dietrich. Hanover: Dartmouth College P, 2012. 68–93. Print.

Daston, Lorraine, and Peter Galison. *Objectivity*. New York: Zone, 2007. Print.

De Man, Paul. *Blindness and Insight: Essays in the Rhetoric of Contemporary Criticism*. 2nd ed. Minneapolis: U of Minnesota P, 1973. Print.

Deleuze, Gilles. *Cinema 2: The Time-Image*. Trans. Hugh Tomlinson and Robert Galeta. Minneapolis: U of Minnesota P, 1989. Print.

—. *Foucault*. Trans. Sean Hand. Minneapolis: U of Minnesota P, 1988. Print.

Deleuze, Gilles, and Felix Guattari. *A Thousand Plateaus*. Trans. Brian Massumi. Minneapolis: U of Minnesota P, 1987. Print.

Deluca, Kevin Michael. *Image Politics: the New Rhetoric of Environmental Activism*. London: Routledge, 1999. Print.

Dennis, Michael Aaron. "Graphic Understanding: Instruments and Interpretation in Robert Hooke's *Micrographia*." *Science in Context* 3.2 (1989): 309–64. Print.

Derrida, Jacques. *Of Grammatology*. Trans. Gayatri Chakravorty Spivak. Baltimore: The Johns Hopkins UP, 1974. Print.

—. "Signature, Event, Context." *Margins of Philosophy*. Trans. Alan Bass. Chicago: U of Chicago P, 1982. 309–30. Print.

Donhauser Z. J., and B. A. Mantooth. "Molecular Switches." Digital image. *Featured Images of the Weiss Group*. The Pennsylvania State University, 17 Aug. 2004. Web. 2 Dec. 2011.

Doyle, Richard. *On Beyond Living: Rhetorical Transformations of the Life Sciences*. Stanford: Stanford UP, 1997. Print.

Drexler, K. Eric. "Differential Gear." *Institute for Molecular Manufacturing*. Institute for Molecular Manufacturing, 3 Oct. 2003. Web. 29 July 2013.

—. *Engines of Creation: The Coming Era of Nanotechnology*. New York: Anchor Press, 1986. Print.

Drucker, Johanna. "Reading Interface." *PMLA: Publications of the Modern Language Association of America* 128.1 (2013): 213–20. Print.

—. *Speclab: Digital Aesthetics and Projects in Speculative Computing*. Chicago: U of Chicago P, 2009. Print.

Dumit, Joseph. *Picturing Personhood: Brain Scans and Biomedical Identity*. Princeton: Princeton UP, 2004. Print.

Edgerton, Samuel Y. *Heritage of Giotto's Symmetry: Art and Science on the Eve of the Scientific Revolution*. Ithaca: Cornell UP, 1991. Print.

Edie, James. *Speaking and Meaning: The Phenomenology of Language*. Bloomington: Indiana UP, 1976. Print.

Edwards, Paul. *The Closed World: Computers and the Politics of Discourse in Cold War America*. Cambridge: MIT P, 1997. Print.

Ehses, Hanno H. J. "Representing *Macbeth*: A Case Study in Visual Rhetoric." *Visual Rhetoric in a Digital World*. Ed. Carolyn Handa. Boston: Bedford/St. Martin's, 2004. 164–76. Print.

Eigler, Don. "From the Bottom Up: Building Things with Atoms." *Nanotechnology*. Ed. Gregory Timp. New York: Springer-Verlag, 1999. 425–36. Print.

Eigler, D. M., and E. K. Schweizer. "Positioning Single Atoms with a Scanning Tunneling Microscope." *Nature* 344 (1990): 524–26. Print.

Eisenstein, Elizabeth. *The Printing Press as an Agent of Change*. London: Cambridge UP, 1979. Print.

Elkins, James. "Art History and the Criticism of Computer-Generated Images." *Leonardo* 27.4 (1994): 335–42. Print.

—. *Domain of Images*. Ithaca: Cornell UP, 1999. Print.

—. "Logic and Images in Art History." *Perspectives on Science* 7.2 (1999): 151–80. Print.

—. *The Object Stares Back: On the Nature of Seeing*. New York: Simon and Schuster, 1995. Print.

—. *Poetics of Perspective*. Ithaca: Cornell UP, 1994. Print.

Engelbart, Douglas. "The Augmented Knowledge Workshop." *A History of Personal Workstations*. Ed. Adele Goldberg. New York: ACM Press, 1988. 188–89. Print.

Faber, Brenton. "Popularizing Nanoscience: The Public Rhetoric of Nanotechnology, 1986–1999." *Technical Communication Quarterly* 15.2 (2006): 141–69. Print.

Fahnestock, Jeanne. "Rhetoric of Science: Enriching the Discipline." *Technical Communication Quarterly* 14.3 (2005): 277–86. Print.

—. *Rhetorical Figures in Science*. Oxford: Oxford UP, 1999. Print.

Feynman, Richard. "There's Plenty of Room at the Bottom." *Engineering and Science* 23.5 (1960): 22-36. Rpt. in *Miniaturization*. Ed. H. D. Gilbert. New York: Reinhold, 1960. 282-96. Print.

Fiedeler, Ulrich. "Technology Assessment of Nanotechnology: Problems and Methods of Assessing Emerging Technologies." *The Yearbook of Nanotechnology in Society, Volume 1: Presenting Futures*. Ed. Erik Fisher, Cynthia Selin, and Jameson M. Wetmore. New York: Springer, 2008. 241–63. Print.

Fisher, Arthur. "Seeing Atoms." *Popular Science* Apr. 1989: 106. Print.

Fisher, Jennifer. "Relational Sense: Towards a Haptic Aesthetics." *Parachute* 87 (1997): 4–11. Print.

Ford, Brian. *Images of Science: A History of Scientific Illustration*. New York: Oxford UP, 1993. Print.

Foster, Hal, ed. *Vision and Visuality*. Seattle: Bay Press, 1988. Print.

Foster, John. "Atomic Imaging and Positioning." *Nanotechnology: Research and Perspectives. Papers from the First Foresight Conference on Nanotechnology*. Ed. B. C. Crandall and James Lewis. Cambridge: MIT P, 1992. 15–36. Print.

Frankel, Felice. "Capturing Quantum Corrals." *American Scientist* 93.3 (2005): 261. Print.

Galison, Peter. "Computer Simulations and the Trading Zone." *The Disunity of Science: Boundaries, Contexts, and Power*. Ed. Peter Galison and David Stump. Palo Alto: Stanford UP, 1996. 118–57. Print.

—. *Image and Logic: A Material Culture of Microphysics*. Chicago: U of Chicago P, 1997. Print.

Galloway, Alexander. *The Interface Effect*. New York: Polity Press, 2012. Print.

Garfinkel, Simson. "Critique of Nanotechnology: A Debate in Four Parts." *Whole Earth Review* 67 (1990): 104–13. Print.

Giaver, Ivan. "Energy Gap in Superconductors Measured By Electron Tunneling." *Physical Review Letters* 5.4 (1960): 147–48. Print.

Gibson, James. *The Ecological Approach to Visual Perception*. Boston: Houghton Mifflin, 1979. Print.

Gimzewski, James K., and Christian Joachim. "Nanoscale Science of Single Molecules Using Local Probes." *Science* 283 (1999): 1683. Print.

Gimzewski, J. K., T. A. Jung, M. T. Cuberes, and R. R. Schlittler. "Scanning Tunneling Microscopy of Individual Molecules: Beyond Imaging." *Surface Science* 386 (1997): 101. Print.

Gimzewski, James, and Victoria Vesna. "The Nanomeme Syndrome: The Blurring of Fact and Fiction in the Construction of a New Science." *Technoetic Arts* 1.1 (2003): 7–24. Print.

Golan, Tal. "The Emergence of the Silent Witness: The Legal and Medical Reception of X-rays in the USA." *Social Studies of Science* 34.4 (2004): 469–99. Print.

Granek, Galina, and Giora Hon. "Searching for Asses, Finding a Kingdom: The Story of the Invention of the Scanning Tunneling Microscope (STM)." *Annals of Science* 65.1 (Jan 2008): 101–25. Print.

Graves, Heather Brodie. *Rhetoric In(to) Science: Style as Invention in Inquiry*. Cresskill: Hampton Press, 2005. Print.

Grey, Francois. "STM-Based Nanotechnology: The Japanese Challenge." *Advanced Materials* 5.10 (1993): 704–10. Print.

Groggin, Maureen Daly. "Visual Rhetoric in Pens of Steel and Inks of Silk: Challenging the Great Visual/Verbal Divide." Hill and Helmers 87–110.

Gross, Alan. "Darwin's Diagram: Scientific Visions and Scientific Visuals." *Ways of Seeing, Ways of Speaking: The Integration of Rhetoric and Vision in Constructing the Real*. Ed. K. Fleckenstein, S. Hum, and L.T. Calendrillo. West Lafayette: Parlor Press, 2007. 52–80. Print.

—. "Toward a Theory of Visual-Verbal Interaction: The Example of Lavoisier." *Rhetoric Society Quarterly* 39.2 (2009): 147–69. Print.

Hacking, Ian. *Representing and Intervening*. Cambridge: Cambridge UP, 1983. Print.

Hallyn, Fernand. "Atoms and Letters." *Metaphor and Analogy in the Sciences.* Ed. Fernand Hallyn. Dordrecht: Kluwer Academic Publishers, 2000. 53–69. Print.

Hansen, Mark B. N. *New Philosophy for New Media.* Cambridge: MIT P, 2004. Print.

—. *Embodying Technesis: Technology Beyond Writing.* Ann Arbor: U of Michigan P, 2000. Print.

Hanson, Valerie L. "Envisioning Ethical Nanotechnology: The Rhetorical Role of Visions in Postponing Societal and Ethical Implications Research." *Science As Culture* 20.1 (2011): 1–36. Print.

Haraway, Donna. *Simians, Cyborgs and Women: The Reinvention of Nature.* New York: Routledge, 1991. Print.

Harrison, Claire. "Visual Social Semiotics: Understanding How Still Images Make Meaning." *Technical Communication* 50.1 (2003): 46–60. Print.

Hayles, N. Katherine. *How We Became Posthuman: Virtual Bodies in Cybernetics, Literature, and Informatics.* Chicago: U of Chicago P, 1999. Print.

—. *How We Think: Digital Media and Contemporary Technogenesis.* Chicago: U of Chicago P, 2012. Print.

—. *Writing Machines.* Cambridge: MIT P, 2002. Print.

Heller, Eric. "Condensed-Matter Physics: Electrons in the Looking Glass." *Nature* 403 (2000): 489–91. Print.

Helmreich, Stefan. *Silicon Second Nature: Culturing Artificial Life in a Digital World.* Berkeley: U of California P, 1998. Print.

Hennig, Jochen. "Changes in the Design of Scanning Tunneling Microscopic Images from 1980 to 1990." *Techne: Research in Philosophy and Technology* 11.1 (2007). Web. 21 Nov. 2013. Rpt. in *Nanotechnology Challenges: Implications for Philosophy, Ethics and Society.* Ed. Joachim Schummer and Davis Baird. Singapore: World Scientific Publishing, 2006. 143–63. Print.

—. "Images and Graphs as Representations of Scanning Probe Microscopic Measurements." Imaging and Imagining Nanoscience and Engineering: An International and Interdisciplinary Conference. Adam's Mark Hotel, Columbia, SC. 4 Mar. 2004. Presentation.

Hesse, Mary. *Revolutions and Reconstructions in the Philosophy of Science.* Brighton, Sussex: Harvester Press, 1980. Print.

Hill, Charles A., and Marguerite Helmers. Introduction. *Defining Visual Rhetorics.* By Charles A. Hill and Marguerite Helmers, eds. Mahwah: Lawrence Erlbaum, 2004. 1–24. Print.

Hooke, Robert. *Micrographia: or Some Physiological Descriptions of Minute Bodies Made by Magnifying Glasses. With Observations and Inquiries Thereupon.* London: Jo. Martyn, 1665. Print.

Hullman, Jessica, and Nicholas Diakopoulos. "Visualization Rhetoric: Framing Effects in Narrative Visualization." *IEEE Transactions on Visualization and Computer Graphics* 17.12 (2011): 2231–40. Print.

IBM. "Corral Reef." *STM Image Gallery*. IBM Research, n.d. Web. 30 July 2013.

IBM. "The Beginning." *STM Image Gallery*. IBM Research, n.d. Web. 30 July 2013.

Ihde, Don. "Hermeneutics and the New Imaging." *Ways of Seeing, Ways of Speaking*. Ed. Kristie S. Fleckenstein, Sue Hum, and Linda T. Calendrillo. West Lafayette: Parlor Press, 2007. 33–51. Print.

Illich, Ivan. "Guarding the Eye in the Age of Show." *RES: Anthropology and Aesthetics* 28 (1995): 47-61. Print.

—. "The Scopic Past and the Ethics of the Gaze: A Plea for the Historical Study of Ocular Perception." 1-17. *Ivan Illich*. www.altraofficina.it. Web. 29 July 2014.

Ippolito, Frank. "The Subtle Beauty of Art in the Service of Science." *Nature* 422 (2003): 15. Print.

Isocrates. *Antidosis, Against the Sophists*. Trans. George Norlin. New York: G. P. Putnam's Sons, 1929. Print.

Ivins, William. *Prints and Visual Communication*. Cambridge: MIT P, 1969. Print.

Jaklevic, R. C. "Real-time Images of Surface Diffusion." *Nature* 331 (1988): 659. Print.

Jarratt, Susan. *Rereading the Sophists: Classical Rhetoric Refigured*. Carbondale: Southern Illinois UP, 1991. Print.

Jay, Martin. *Downcast Eyes: The Denigration of Vision in Twentieth-Century French Thought*. Berkeley: U of California P, 1994. Print.

Joy, Bill. "Why the Future Doesn't Need Us." *Wired* 8.04 (April 2000): 238–62. Print.

Joyce, Kelly A. *Magnetic Appeal: MRI and the Myth of Transparency*. Ithaca: Cornell UP, 2008. Print.

Joyce, Michael. *Of Two Minds: Hypertext, Pedagogy and Poetics*. Ann Arbor: U of Michigan P, 1996. Print.

Kaiser, David. *Drawing Theories Apart: The Dispersion of Feynman Diagrams in Postwar Physics*. Chicago: U of Chicago P, 2005. Print.

Kandel, S. A., and P. S. Weiss. "Binding and Mobility of Atomically Resolved Cobalt Clusters on Molybdenum Disulfide." *Journal of Physical Chemistry B* 105 (2001): 8103. Print.

Kaplan, Sarah, and Joanna Radin. "Bounding an Emerging Technology: Para-Scientific Media and the Drexler-Smalley Debate about Nanotechnology." *Social Studies of Science* 41.4 (2011): 457–85. Print.

Keller, Evelyn Fox. "The Biological Gaze." *FutureNatural: Nature, Science, Culture*. Ed. George Robertson, Melinda Mash, Lisa Tickner, Jon Bird,

Barry Curtis, and Tim Putnam. New York: Routledge, 1996. 107–21. Print.

—. *Making Sense of Life: Explaining Biological Development with Models, Metaphors, and Machines*. Cambridge: Harvard UP, 2002. Print.

—. *Refiguring Life: Metaphors of Twentieth-Century Biology*. New York: Columbia UP, 1995. Print.

Kenney, Keith. "Building Visual Communication Theory by Borrowing from Rhetoric." *Visual Rhetoric in a Digital World*. Ed. Carolyn Handa. Boston: Bedford/St. Martin's, 2004. 321–43. Print.

Kerferd, G. E. *The Sophistic Movement*. Cambridge: Cambridge UP, 1981. Print.

Khalip, Jacques, and Robert Mitchell, eds. *Releasing the Image: From Literature to New Media*. Palo Alto: Stanford UP, 2011. Print.

Kirschenbaum, Matthew. *Mechanisms: New Media and the Forensic Imagination*. Cambridge: MIT P, 2008. Print.

Kittler, Friedrich. "The Perspective of Print." Trans. Geoffrey Winthrop-Young and Michael Wutz. *Configurations* 10 (2002): 37–50. Print.

Kostelnick, Charles. "Melting-Pot Ideology, Modernist Aesthetics and the Emergence of Graphical Conventions: The Statistical Atlases of the United States, 1874–1925." Hill and Helmers 215–42.

Kostelnick, Charles, and David D. Roberts. *Designing Visual Language: Strategies for Professional Communicators*. Needham Heights: Allyn and Bacon, 1998. Print.

Kostelnick, Charles, and Michael Hassett. *Shaping Information: The Rhetoric of Visual Conventions*. Carbondale: Southern Illinois UP, 2003. Print.

Kress, Gunther, and Theo Van Leeuwen. *Reading Images: The Grammar of Visual Design*. London: Routledge, 1996. Print.

Kushmerick, J. G., S. A. Kandel, P. Han, J. A. Johnson, and P. S. Weiss. "Atomic-Scale Insights into Hydrodesulfurization." *Journal of Physical Chemistry B* 104 (2000): 2980–88. Print.

Lakoff, George, and Mark Johnson. *Metaphors We Live By*. Chicago: U of Chicago P, 1980. Print.

Landow, George P. "The Rhetoric of Hypermedia: Some Rules for Authors." *Journal of Computing in Higher Education* 1.1 (1989): 39–64. Print.

Lant, Antonia. "Haptical Cinema." *October* 75 (1995): 45–73. Print.

Latour, Bruno. "Drawing Things Together." Lynch and Woolgar 19–68.

—. *Science in Action: How to Follow Scientists and Engineers through Society*. Cambridge: Harvard UP, 1987. Print.

Latour, Bruno, and Steve Woolgar. *Laboratory Life: The Construction of Scientific Facts*. Princeton: Princeton UP, 1986. Print.

Leiner, Dominik J., and Olivier Quiring. "What Interactivity Means to the User: Essential Insights into and a Scale for Perceived Interactivity." *Journal of Computer-Mediated Communication* 14 (2008): 127–55. Print.

Lenoir, Timothy. *Instituting Science: The Cultural Production of Scientific Disciplines*. Stanford: Stanford UP, 1997. Print.

—. "Shaping Biomedicine as an Information Science." *Proceedings of the 1998 Conference on the History and Heritage of Science Information Systems*. Ed. Mary Ellen Bowden, Trudi Bellardo Hahn, and Robert V. Williams. Medford: Information Today, 1998. 27–45. Print.

Libet, Benjamin. *Neurophysiology of Consciousness: Selected Papers and New Essays*. Boston: Birkhåuser, 1993. Print.

Lindberg, David C. *Theories of Vision from Al-Kindi to Kepler*. Chicago: U of Chicago P, 1976. Print.

Lopez, Jose Julian. "Nanotechnology: Legitimacy, Narrative, and Emergent Technologies." *Sociology Compass* 2.4 (2008): 1266–86. Print.

—. "Notes on Metaphors, Notes as Metaphors: the Genome as Musical Spectacle." *Science Communication* 29.1 (2007): 7–34. Print.

Lösch, Andreas. "Anticipating the Futures of Nanotechnology: Visionary Images as Means of Communication." *Technology Analysis and Strategic Management* 18.3/4 (2006): 393–409. Print.

Lovejoy. Alan. "Re: Proto-assemblers Anyone?" *Sci.Nanotech Archives*. Foresight.org, 8 Apr. 1990. Web. 26 Mar. 2003.

Lupton, Ellen, and J. Abbott Miller. *Design Writing Research*. New York: Princeton Architectural Press, 1996. Print.

Lynch, Michael. *Art and Artifact in Laboratory Science: A Study of Shop Work and Shop Talk in a Research Laboratory*. London: Routledge, 1985. Print.

—. "The Externalized Retina: Selection and Mathematization in the Visual Documentation of Objects in the Life Sciences." Lynch and Woolgar 153–86.

—. "Laboratory Space and the Technological Complex: An Investigation of Topical Contextures." *Science in Context* 4.1 (1991): 51–78. Print.

Lynch, Michael, and Samuel Y. Edgerton, Jr. "Aesthetics and Digital Image Processing: Representational Craft in Contemporary Astronomy." *Picturing Power: Visual Depiction and Social Relations*. Ed. Gordon Fyfe and John Law. London: Routledge, 1988. 184–220. Print.

Lynch, Michael, and Steve Woolgar, eds. *Representation in Scientific Practice*. Cambridge: MIT P, 1990. Print.

Manovich, Lev. *The Language of New Media*. Cambridge: MIT P, 2000. Print.

Mantooth, Brent. "Nanoscopic Analyses: Single Molecule Characterization in Molecular Electronics and Surface Science." Diss. The Pennsylvania State University, 2005. Print.

Marks, Laura U. *The Skin of the Film: Intercultural Cinema, Embodiment and the Senses*. Durham: Duke UP, 2000. Print.

—. *Touch: Sensuous Theory and Multisensory Media*. Minneapolis: U of Minnesota P, 2002. Print.

Marrou, H. I. *A History of Education in Antiquity.* Trans. George Lamb. New York: Sheed and Ward, 1956. Print.

Massumi, Brian. *Parables for the Virtual: Movement, Affect, Sensation.* Durham: Duke UP, 2002. Print.

McCray, W. P. "Will Small Be Beautiful? Making Policies for Our Nanotech Future." *History and Technology* 21.2 (2005): 177–203. Print.

McMillan, Sally J. "Exploring Models of Interactivity from Multiple Research Traditions: Users, Documents, and Systems." *Handbook of New Media: Social Shaping and Consequences of ICTs.* Ed. Leah A. Lievrouw and Sonia Livingstone. London: SAGE Publications, 2003. 163–83. Print.

Merz, Martina. "Designed for Travel: Communicating Facts Through Images." *How Well Do Facts Travel? The Dissemination of Reliable Knowledge.* Ed. Peter Howlett and Mary S. Morgan. Cambridge: Cambridge UP, 2010. 349–75. Print.

Milburn, Colin. "Atoms and Avatars: Virtual Worlds as Massively-Multiplayer Laboratories." *Spontaneous Generations* 2.1 (2008): 63–89. Print.

—. "Digital Matters: Video Games and the Cultural Transcoding of Nanotechnology." *Governing Future Technologies: Nanotechnology and the Rise of an Assessment Regime.* Ed. Mario Kaiser, Monika Kurath, Sabine Maasen, and Christoph Rehmann-Sutter. Dordrecht: Springer, 2009. 109–27. Print.

—. *Nanovision: Engineering the Future.* Durham: Duke UP, 2008. Print.

Miller, Arthur I. "Metaphor and Scientific Creativity." *Metaphor and Analogy in the Sciences.* Ed. Fernand Hallyn. Dordrecht: Kluwer Academic Publishers, 2000. 147–64. Print.

Miller, Carolyn R. "Foreword: Rhetoric, Technology, and the Pushmi-Pullyu." *Rhetorics and Technologies: New Directions in Writing and Communication.* Ed. Stuart Selber. Columbia: U of South Carolina P, 2010. ix-xii. Print.

Mirkin, Chad. "Nanotechnology: Fact or Fiction." *Chemical and Engineering News* (2001): 185. Print.

Mody, Cyrus C. M. *Instrumental Community: Probe Microscopy and the Path to Nanotechnology.* Cambridge: MIT P, 2011. Print.

—. "Intervening Technology, Representing Technique: Probe Microscopy and the Art of the Nanoworld." Imaging and Imagining Nanoscience and Engineering: An International and Interdisciplinary Conference. Adam's Mark Hotel, Columbia, SC. 4 Mar. 2004. Presentation.

Mody, Cyrus C. M., and Michael Lynch. "Test Objects and Other Epistemic Things: A History of A Nanoscale Object." *British Journal of the History of Science* 43.3 (2010): 423–58. Print.

Mulhall, Douglas. *Our Molecular Future: How Nanotechnology, Robotics, Genetics and Artificial Intelligence Will Transform Our World.* Amherst: Prometheus, 2002. Print.

Müller, Erwin W. "Resolution of the Atomic Structure of a Metal Surface by the Field Ion Microscope." *Journal of Applied Physics* 27 (1956): 474–77. Print.

Müller, Erwin W., and Kanwar Bahadur. "Field Ionization of Gases at a Metal Surface and the Resolution of the Field Ion Microscope." *Physical Review* 102.3 (1956): 624–31. Print.

Munster, Anna. *Materializing New Media: Embodiment in Information Aesthetics.* Hanover: Dartmouth College P, 2006. Print.

Murray, Janet Horowitz. *Hamlet on the Holodeck: The Future of Narrative in Cyberspace.* New York: Free Press, 1997. Print.

Myers, Greg. "Every Picture Tells a Story: Illustrations in E. O. Wilson's *Sociobiology.*" Lynch and Woolgar 231–66.

Myers, Natasha. "Animating Mechanism: Animations and the Propagation of Affect in the Lively Arts of Protein Modelling." *Science Studies* 19.2 (2006): 6–30. Print.

"The Nanomanipulator." *warrenrobinett.com.* Warren Robinett, n.d.. Web. 17 July 2013.

Nelson, Robert S. Introduction. *Visuality Before and Beyond the Renaissance: Seeing as Others Saw.* By Robert S. Nelson, ed. Cambridge: Cambridge UP, 2000. 1–21. Print.

Nerlich, Brigitte. "Powered by Imagination: Nanobots at the Science Photo Library." *Science as Culture* 17 (2008): 269–92. Print.

Nerlich, Brigitte, Richard Elliott, and Brendon Larson, eds. *Communicating Biological Sciences: Ethical and Metaphorical Dimensions.* Farnham: Ashgate Publishing, 2009. Print.

Nishizaka, Aug. "The Embodied Organization of a Real-time Fetus: The Visible and the Invisible in Prenatal Ultrasound Examinations." *Social Studies of Science* 41.3 (2011): 309–36. Print.

Nordmann, Alfred. "Nanotechnology's Worldview: New Space for Old Cosmologies." *IEEE Technology and Society Magazine* 23 (2004): 48–54. Print.

OhAnluain, Daithi. "Reaching Through the Net to Touch." *Wired* 3 July 2003. Web. 29 July 2013.

Ortony, Andrew, ed. *Metaphor and Thought.* 2nd ed. Cambridge: Cambridge UP, 1993. Print.

Ottino, Julio M. "Is A Picture Worth 1,000 Words? Exciting New Illustration Technologies Should Be Used With Care." *Nature* 421 (2003): 476. Print.

—. "The Role of Images in Science and Engineering." Imaging and Imagining Nanoscience and Engineering: An International and Interdisciplinary Conference. Adam's Mark Hotel, Columbia, SC. 5 Mar. 2004. Presentation.

Oudshoorn, Nelly. *Beyond the Natural Body: An Archaelogy of Sex Hormones.* London: Routledge, 1994. Print.

Oveido, Ollie, Joyce R. Walker, and Byron Hawk, eds. *Digital Tools in Composition Studies: Critical Dimensions and Implications.* Cresskill: Hampton Press, 2010. Print.

Oyama, Susan. *The Ontogeny of Information: Developmental Systems and Evolution.* 2nd ed. Durham: Duke UP, 2000. Print.

Park, Sang-Il, and Robert C. Barrett. "Design Considerations for an STM System." *Scanning Tunneling Microscopy.* Ed. Joseph Stroscio and William Kaiser. San Diego: Academic Press, 1993. 31–75. Print.

Parton, Nigel. "Changes in the Form of Knowledge in Social Work: From the 'Social' to the 'Informational'?" *British Journal of Social Work* 38 (2008): 253–69. Print.

Pascual, J. L., J. J. Jackiw, K. F. Kelly, H. Conrad, H. P. Rust, and P. S. Weiss. "Local Electronic Structural Effects and Measurements on the Adsorption of Benzene on Ag{110}." *Physical Review B* 62.19 (2000): 12632–35. Print.

Pauwels, Luc, ed. *Visual Cultures of Science: Rethinking Representational Practices in Knowledge Building and Science Communication.* Hanover: Dartmouth College P, 2006. Print.

Poulakos, Takis. *Speaking for the Polis.* Columbia: South Carolina UP, 1997. Print.

Porter, James. "Recovering Delivery for Digital Rhetoric." *Computers and Composition* 26 (2009): 207–24. Print.

Rafaeli, Sheizaf, and Fay Sudweeks. "Networked Interactivity." *Journal of Computer-Mediated Communication* 2.4 (1997). Web. 29 July 2013.

Rasmussen, Nicolas. *Picture Control: The Electron Microscope and the Transformation of Biology in America, 1940–1960.* Stanford: Stanford UP, 1997. Print.

Regis, Ed. *Nano: The Emerging Science of Nanotechnology: Remaking the World—Molecule by Molecule.* Boston: Little, 1995. Print.

Riegl, Alois. *Late Roman Art Industry.* 1902. Trans. Rolf Winkes. Rome: Giorgio Bretschneider Editore, 1985. Print.

Rice, Jeff. "Occupying the Digital Humanities." *College English* 75.4 (2013): 360–78. Print.

Richards, I. A. *The Philosophy of Rhetoric.* London: Oxford UP, 1936. Print.

Rip, Arie, and Ruivenkamp, Martin. "Visualizing the Invisible Nanoscale Study: Visualization Practices in Nanotechnology Community of Practice." *Science Studies* 23.1 (2010): 3–36. Print.

Robinson, Ken. "The Book of Nature." *Into Another Mould: Change and Continuity in English Culture 1625–1700*. Ed. T. G. S. Cain and Ken Robinson. London: Routledge, 1992. 86–106. Print.

Romanyshyn, Robert D. "The Despotic Eye and Its Shadow: Media Image in the Age of Literacy." *Modernity and the Hegemony of Vision*. Ed. David Michael Levin. Berkeley: U of California P, 1993. 339–60. Print.

Roskies, Adina L. "Neuroimages, Pedagogy, and Society." *The Educated Eye: Visual Culture and Pedagogy in the Life Sciences*. Ed. Nancy Anderson and Michael R. Dietrich. Hanover: Dartmouth College P, 2012. 255–76. Print.

Rotman, Brian. *Signifying Nothing: The Semiotics of Zero*. Stanford: Stanford UP, 1987. Print.

Rudwick, Martin J. S. "The Emergence of a Visual Language for Geological Science 1760–1840." *History of Science* 14 (1976): 149–95. Print.

Ruse, Michael. "Are Pictures Really Necessary? The Case of Sewall Wright's 'Adaptive Landscapes.'" *PSA* (1990): 63–77.

Russ, John. *Image Processing Handbook*. 6th ed. Boca Raton: CRC Press, 2011. Print.

Russell, Philip E. "Scanned Probe Microscopy: Overview and Applications of Critical Tools for Nanotechnology." Imaging and Imagining Nanoscience and Engineering: An International and Interdisciplinary Conference. Adam's Mark Hotel, Columbia, SC. 7 Mar. 2004. Presentation.

Rystedt, Hans, Jonas Ivarsson, Sara Asplund, Ase Allansdotter Johnsson, and Magnus Bath. "Rediscovering Radiology: New Technologies and Remedial Action at the Worksite." *Social Studies of Science* 41.6 (2011): 867–91. Print.

Salinas, Carlos. "Technical Rhetoricians and the Art of Configuring Images." *Technical Communication Quarterly* 11.2 (2002): 165–83. Print.

Salling, C. T., and M. G. Lagally. "Fabrication of Atomic-Scale Structures on Si(001) Surfaces." *Science* 265 (1994): 503. Print.

Sauer, Beverly. *Rhetoric of Risk: Technical Documentation in Hazardous Environments*. Mahwah: Lawrence Earlbaum, 2003. Print.

Schriver, Karen. *Dynamics in Document Design: Creating Text for Readers*. New York: Wiley, 1996. Print.

Selber, Stuart. Introduction. *Rhetorics and Technologies: New Directions in Writing and Communication*. By Stuart Selber, ed. Columbia: U of South Carolina P, 2010. 1–14. Print.

Selin, Cynthia. "Expectations and the Emergence of Nanotechnology." *Science, Technology, and Human Values* 32.2 (2007): 196–220. Print.

Shannon, Claude E., and Warren Weaver. *The Mathematical Theory of Communication*. Urbana: U of Illinois P, 1949. Print.

Shapin, Steven, and Simon Schaffer. *Leviathan and the Air Pump: Hobbes, Boyle, and the Experimental Life*. Princeton: Princeton UP, 1985. Print.

Shaviro, Steven. *The Cinematic Body*. Minneapolis: U of Minnesota P, 1993. Print.

Shew, Ashley. "Nanotech's History: An Interesting, Interdisciplinary, Ideological Split." *Bulletin of Science, Technology, and Society* 28.5 (2008): 390–99. Print.

Sidler, Michelle. "The Rhetoric of Cells: Understanding Molecular Biology in the Twenty-First Century." *Rhetoric Review* 25.1 (2006): 58–75. Print.

Snyder, Joel. "Picturing Vision." *The Language of Images*. Ed. W. J. T. Mitchell. Chicago: U of Chicago P, 1974–1980. 219–46. Print.

Stafford, Barbara Maria. *Artful Science: Enlightenment, Entertainment, and the Eclipse of Visual Education*. Cambridge: MIT P, 1994. Print.

Stengel, Richard. "Best of '90: Well, Hello to '90s Humility." *Time* 31 Dec. 1990: 40–42. Print.

Stewart, Susan. *On Longing: Narratives of the Miniature, the Gigantic, the Souvenir, the Collection*. Durham: Duke UP, 1993. Print.

Stoll, E. P. "Picture Processing and Three-Dimensional Visualization of Data from Scanning Tunneling and Atomic Force Microscopy." *IBM Journal of Research and Development* 35.1/2 (1991): 67–77. Print.

Stroscio, Joseph A., and D. M. Eigler. "Atomic and Molecular Manipulation with the Scanning Tunneling Microscope." *Science* 254 (1991): 1319–26. Print.

Stroscio, Joseph, and R. M. Feenstra. "Methods of Scanning Tunneling Microscopy." *Scanning Tunneling Microscopy*. Ed. Joseph Stroscio and William Kaiser. San Diego: Academic Press, 1993. 96–145. Print.

Sutter, P., P. Zahl, E. Sutter, and J. E. Bernard. "Energy-Filtered Scanning Tunneling Microscopy using a Semiconductor Tip." *Physical Review Letters* 90 (2003): 166101. Print.

Taussig, Michael. *Mimesis and Alterity: A Particular History of the Senses*. New York: Routledge, 1993. Print.

Taylor, Russell M., Warren Robinett, Vernon L. Chi, Frederick P. Brooks, Jr., William V. Wright, R. Stanley Williams, and Erik J. Snyder. "The Nanomanipulator: A Virtual-Reality Interface for a Scanning Tunneling Microscope." *Computer Graphics: Proceedings of SIGGRAPH '93* (1993): 127–34. Print.

Thomas, Rosalind. *Orality and Literacy in Ancient Greece*. Cambridge: Cambridge UP, 1992. Print.

Timp, Gregory, ed. *Nanotechnology*. New York: Springer-Verlag, 1999. Print.

Toumey, Chris. "Apostolic Succession: Does Nanotechnology Descend from Richard Feynman's 1959 Talk?" *Engineering and Science* 68.1 (2005):16–23. Print.

Tufte, Edward. *Envisioning Information*. Cheshire: Graphics Press, 1990. Print.

United States. National Nanotechnology Initiative. "NNI Budget." *Nano.gov.* National Nanotechnology Initiative, n.d. Web. 30 July 2013.

United States. National Nanotechnology Initiative. "What It Is and How It Works." *Nano.gov.* National Nanotechnology Initiative, n.d. Web. 30 July 2013.

United States. National Science and Technology Council. Committee on Technology. *Nanotechnology: Shaping the World Atom by Atom.* Washington: GPO, 1999. Web. 29 July 2014.

United States. Subcommittee on Nanoscale Science, Engineering, and Technology, Committee on Technology, and National Science and Technology Council. *The National Nanotechnology Initiative: Research and Development Leading to a Revolution in Technology and Industry: Supplement to the President's 2011 Budget.* Washington: GPO, 2010. Web. 8 July 2010.

United States. Subcommittee on Nanoscale Science, Engineering, and Technology, Committee on Technology, and National Science and Technology Council. *The National Nanotechnology Initiative. Supplement to the President's Budget for Fiscal Year 2014.* Washington: GPO, 2013. Web. 29 July 2013.

Van Dyck, Robert, Philip Ekstrom, and Hans Dehmelt. "Axial, Magnetron, and Spin-Cyclotron Beat Frequencies Measured on Single Electron Almost at Rest in Free Space (Geonium)." *Nature* 262 (1976): 776. Print.

van Eck, Caroline. *Classical Rhetoric and the Visual Arts in Early Modern Europe.* New York: Cambridge UP, 2007. Print.

Van Hook, LaRue. "Alcidamas Versus Isocrates: The Spoken Versus the Written Word." *Classical Weekly* 12:12 (1919): 89–94. Print.

"Veeco SPM Calendar Image Competition—Deadline is 2 October." *Microscopy and Analysis.* Web. 17 July 2013.

Vivian, Bradford. "In the Regard of the Image." *JAC: Journal of Advanced Composition* 27.3–4 (2007): 471–504. Print.

Von Baeyer, Hans Christian. *Taming the Atom: The Emergence of the Visible Microworld.* New York: Random House, 1992. Print.

Wardrip-Fruin, Noah, and Nick Montfort. *The New Media Reader.* Cambridge: MIT P, 2003. Print.

Wartofsky, Marx. "Picturing and Representing." *Perception and Pictorial Representation.* Ed. Calvin F. Nodine and Dennis F. Fisher. New York: Praeger, 1979. 272–83. Print.

Welch, Kathleen. *Electric Rhetoric: Classical Rhetoric, Oralism, and a New Literacy.* Cambridge: MIT P, 1999. Print.

Wickman, Chad. "Observing Inscriptions at Work: Visualization and Text Production in Experimental Physics Research." *Technical Communication Quarterly* 22 (2013): 150–71. Print.

—. "Rhetoric, Techne, and the Art of Scientific Inquiry." *Rhetoric Review* 31.1 (2012): 21–40. Print.

Wickramasinghe, H. Kumar. "Extensions of STM." *Scanning Tunneling Microscopy.* Ed. Joseph Stroscio and William Kaiser. San Diego: Academic Press, 1993. 77–93. Print.

Winner, Langdon. *The Whale and the Reactor: A Search for Limits in an Age of High Technology.* Chicago: U of Chicago P, 1986. Print.

Woodruff, D. P. "Is Seeing Believing? Atomic-scale Imaging of Surface Structures Using Scanning Tunnelling Microscopy." *Current Opinion in Solid State and Materials Science* 7 (2003): 75–81. Print.

Woolgar, Steve. "Time and Documents in Researcher Interaction: Some Ways of Making Out What is Happening in Experimental Science." Lynch and Woolgar 123–52.

Worringer, Wilhelm. *Abstraction and Empathy: A Contribution to the Psychology of Style.* Trans. Michael Bullock. New York: International Universities Press, 1963. Print.

Wouters, Paul, Anne Beaulieu, Andrea Scharnhorst, and Sally Wyatt, eds. *Virtual Knowledge: Experimenting in the Humanities and Social Sciences.* Cambridge: MIT P, 2013. Print.

Young, Russell, John Ward, and Fredric Scire. "The Topografiner: An Instrument for Measuring Surface Microtopography." *The Review of Scientific Instruments* 43.7 (1972): 999–1011. Print.

Zwierlein, Anne-Julia. "Queen Mab Under the Microscope: The Invention of Subvisible Worlds in Early Modern Science and Poetry." *Spatial Change in English Literature.* Ed. Joachim Frenk. Trier, Germany: Wissenschaftlicher, 2000. 69–97. Print.

Index

About the Author

Valerie Hanson is Associate Professor of Writing in the College of Science, Health, and the Liberal Arts at Philadelphia University. Her research focuses on the rhetoric of science, visual rhetoric, and rhetoric of technology. She has been a fellow in an interdisciplinary research group at the *Zentrum für interdisziplinäre Forschung* (Center for Interdisciplinary Research) at the University of Bielefeld (Germany). She has published articles on the rhetoric of nanotechnology in journals such as *Science Communication* and *Science as Culture.*

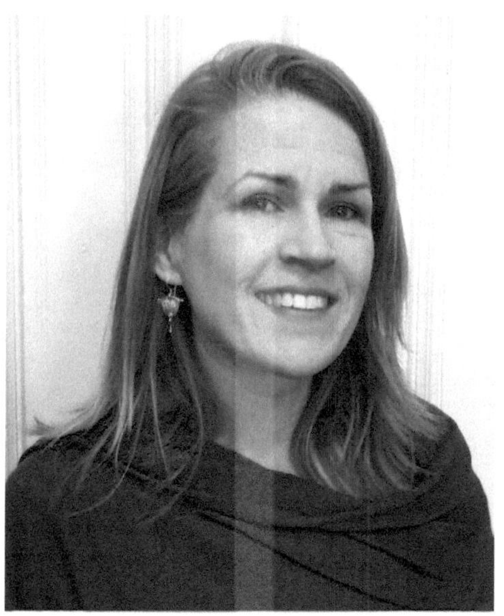

Photograph of the author by Neil Andress.
Used by Permission.

www.ingramcontent.com/pod-product-compliance
Lightning Source LLC
LaVergne TN
LVHW091137080826
845145LV00008B/2177

* 9 7 8 1 6 0 2 3 5 5 5 0 7 *